Advances in Single Molecule, Real-Time (SMRT) Sequencing

Advances in Single Molecule, Real-Time (SMRT) Sequencing

Special Issue Editors

Adam Ameur
Matthew S. Hestand

MDPI • Basel • Beijing • Wuhan • Barcelona • Belgrade

MDPI

Special Issue Editors

Adam Ameur
Uppsala University
Sweden

Matthew S. Hestand
Cincinnati Children's Hospital Medical Center/
University of Cincinnati
USA

Editorial Office
MDPI
St. Alban-Anlage 66
4052 Basel, Switzerland

This is a reprint of articles from the Special Issue published online in the open access journal *Genes* (ISSN 2073-4425) from 2018 to 2019 (available at: https://www.mdpi.com/journal/genes/special_issues/SMRT_sequencing)

For citation purposes, cite each article independently as indicated on the article page online and as indicated below:

LastName, A.A.; LastName, B.B.; LastName, C.C. Article Title. *Journal Name* **Year**, *Article Number*, Page Range.

ISBN 978-3-03921-700-7 (Pbk)
ISBN 978-3-03921-701-4 (PDF)

Contents

About the Special Issue Editors

Adam Ameur (Ph.D.). Dr. Ameur is Associate Professor at the SciLifeLab National Genomics Infrastructure (NGI) in Sweden. He originally trained as a computer scientist and has worked in the field of next-generation sequencing (NGS) for over a decade. In 2008, he received his doctorate degree in bioinformatics from Uppsala University, Sweden, and to date, has authored over 50 scientific publications in genomics and bioinformatics. His current research is focused on technology development and novel sequencing applications for the study of human health and disease. Ongoing activities include the coordination of data analysis within the SweGen project, a national effort that aims to construct a whole genome map of genetic variation in the Swedish population, as well as the introduction of long-read single-molecule sequencing into clinical routines. Since 2017, he has held the position of Adjunct Researcher at the Department of Epidemiology and Preventive Medicine, Monash University, Melbourne, Australia.

Matthew Hestand (Ph.D.). Dr. Hestand is Assistant Professor at the Department of Pediatrics, University of Cincinnati, and Assistant Director of Bioinformatics in the Division of Human Genetics at the Cincinnati Children's Hospital Medical Center. In 2010, he completed his PhD in medicine at Leiden University, situated at the Leiden University Medical Center, the Netherlands, which included implementation of next-generation sequencing (NGS) when the technology was still in its infancy. As a Postdoc at the Gluck Equine Research Center in Kentucky, USA, he then focused on using NGS to refine the equine reference genome and its annotation. From 2012 to 2017, he was instrumental in implementing long-read single-molecule sequencing for a multitude of basic and medical research applications at the Department of Human Genetics, KU Leuven, Belgium. At present, he continues to advance genomics medicine by streamlining and accelerating clinical NGS bioinformatic analyses while implementing new and improved genomic methods, including applications of single-molecule sequencing.

Preface to "Advances in Single Molecule, Real-Time (SMRT) Sequencing"

Since its commercial introduction in 2011, PacBio's single-molecule real-time (SMRT) sequencing technology has transitioned from being a niche platform to a widely adopted tool in basic science, agriculture, and medical research. The primary driver of this transition has been the length of SMRT reads, reaching several tens of kilobases, which allows for more complete assembly and detection of the complex variation in genomic and transcriptomic structures. In addition, SMRT sequencing has the ability to generate high-accuracy reads of single molecules and to detect epigenetic modifications in native DNA—two properties that are missing from the current short-read platform offerings.

In this collection of work, SMRT sequencing has been applied to a broad spectrum of different living organisms, including mushrooms, moths, mosquitos, and humans. This book also addresses new SMRT sequencing methodologies, including phasing of diploid methylation patterns, lower DNA input requirements, and the application of improved short-read re-analysis within a target population. This demonstrates the growing realization across the life sciences that long SMRT read information is required to obtain a more complete view of the genomic, transcriptomic, and/or epigenomic composition of an organism.

Adam Ameur, Matthew S. Hestand
Special Issue Editors

GCAT
TACG
GCAT
genes

MDPI

Editorial

The Versatility of SMRT Sequencing

Matthew S. Hestand [1,2,*] and Adam Ameur [3,4,*]

1   Division of Human Genetics, Cincinnati Children's Hospital Medical Center, Cincinnati, OH 45202, USA
2   Department of Pediatrics, University of Cincinnati College of Medicine, Cincinnati, OH 45202, USA
3   Department of Immunology, Genetics and Pathology, Uppsala University, Science for Life Laboratory, 75025 Uppsala, Sweden
4   Department of Epidemiology and Preventive Medicine, Monash University, Melbourne 32901, Australia
*   Correspondence: matthew.hestand@cchmc.org (M.S.H.); adam.ameur@igp.uu.se (A.A.)

Received: 20 December 2018; Accepted: 3 January 2019; Published: 4 January 2019

The adoption of single molecule real-time (SMRT) sequencing [1] is becoming widespread, not only in basic science, but also in more applied areas such as agricultural, environmental, and medical research. SMRT sequencing offers important advantages over current short-read DNA sequencing technologies, including exceptionally long read lengths (20 kb or more), unparalleled consensus accuracy, and the ability to sequence native, non-amplified, DNA molecules. These sequencing characteristics enable creation of highly accurate de novo genome assemblies, characterization of complex structural variation, direct characterization of nucleotide base modifications, full-length RNA isoform sequencing, phasing of genetic variants, low frequency mutation detection, and clonal evolution determination [2,3]. This Special Issue of Genes is a collection of articles showcasing the latest developments and the breadth of applications enabled by SMRT sequencing technology.

In basic science, SMRT sequencing enables studies into the molecular mechanisms of living cells at a new level of resolution. Perhaps the most advantageous feature of SMRT sequencing is that it facilitates sequencing of long DNA molecules at a very high accuracy. This has enabled the construction of high-quality reference genomes for a wide range of species, including new human genome assemblies, as presented in this special issue [4]. In addition, when SMRT sequencing is performed on native non-amplified DNA molecules, it is possible to access several layers of additional information hidden in the kinetic signals emitted by the polymerases during the sequencing reaction [1]. This kinetic information has been used to detect epigenetic modifications at base pair resolution and even phasing of methylation signatures in diploid organisms, as presented in this special issue [5]. Several important discoveries have already been made from this kinetic information, such as the widespread presence of 6mA modifications in the human genome [6], a modification that was previously thought to only be present in bacterial genomes. In addition to base modifications, SMRT sequencing data also enables us to study other events, such as DNA conformations [7]. Another aspect of SMRT sequencing is that it can be used to study RNA, and it is currently the only technology that can generate high-quality continuous sequences for full-length transcripts up to 10 kb or more. This makes it possible to study splicing variation at a completely new level of resolution [8,9]. SMRT sequencing is also paving the way for a new generation of computational approaches to explore and interpret these rich datasets [10–12]. In summary, SMRT sequencing is enhancing and even opening up new areas of basic research that were not accessible with previous sequencing technologies.

In terms of more applied areas, agriculture is benefiting from the advent of SMRT sequencing for examining important microbes, plants, and animals. SMRT sequencing, often with complementary technologies, has produced new genome assemblies for important crops, such as apples, maize, wine grapes, coffee, rice, black raspberries, asparagus, and cotton [11,13–20]. SMRT transcriptome sequencing has also given new insights into gene structures for rice, wheat, maize, sorghum, barley, and cotton [18,21–25]. Besides providing new references, these projects will improve plant cultivation, such as identifying drought and disease resistant genes. Strategies to detect genetically modified organisms (GMOs) have also been proposed and enhanced with SMRT sequencing [26]. Animal genome assemblies have been produced for several agriculturally valuable species, such as the horse, cow, goat, chicken (including its transcriptome), and commercially important fish like haddock and cod [27–33]. These will lead to improvements in animal breeding, management, and disease resistance. Finally, sequencing of pathogenic bacteria and fungi affecting agriculturally important species is providing insight into the diversity and virulence factors of these pathogens, which in turn will assist in disease risk and management [34–36].

In environmental research, systematic efforts are ongoing to generate reference sequences for thousands of bacterial strains and microorganisms. Recently, this has expanded to the genomes of larger organisms, including vertebrates [37]. SMRT sequencing can also play an important role in ecology research, such as monitoring the composition of fungi in environmental soil or water samples [38,39]. New high-quality references for animal genomes, such as the great apes [40], will provide an invaluable resource for future evolutionary studies. During the last few years, new genome assemblies have also been created for several endangered species, including Hawaii's last crow species [41], aiding in conservation efforts.

Though SMRT sequencing has primarily been applied to basic research, there is a growing implementation for clinical utility [3,42]. The long and highly accurate reads produced from SMRT sequencing have proven to be useful to resolve complex and repetitive regions of the human genome associated with disease. SMRT sequencing is also a sensitive method to detect minor variants in cancer and infectious disease. Although most current methods are based on targeted sequencing, the value of long reads is also becoming apparent for whole-genome sequencing, which allows clinical professionals to resolve repeat expansions, transposable element insertions, and other complex genomic rearrangements that are difficult or even impossible to assess using short-read sequence data.

As we look forward, this technology will provide even longer and more accurate reads at a higher throughput. This will enable routine de novo assembly of both alleles in large diploid genomes, accompanied with tissue specific epigenetic DNA modification information. As a consequence, there will be a demand for a new generation of computational tools to compare complete genomes to each other, as opposed to a reference standard, and to phase genetic variants and epigenetic modifications over large chromosomal regions. By sequencing thousands of individuals with long reads, it will be possible to obtain a detailed picture of complex structural variation within large population cohorts of humans, as well as for other species. Such endeavors will give new insights to the function of the repetitive parts of the genome, and likely explain the cause of many genomic diseases. Looking further on the horizon, SMRT sequencing can be envisioned in combination with other technical advances, such as single cell sequencing to provide information on the epigenetic modifications occurring in single cells. SMRT sequencing has been steadily evolving since the commercial introduction of the technology in 2011. Just as short-read technologies have replaced microarrays and Sanger sequencing for a host of applications, we envision long-read single-molecule sequencing to replace short-read platforms for a majority of applications, as well as continue to evolve into new applications, throughout many different areas in the coming decade.

References

1. Eid, J.; Fehr, A.; Gray, J.; Luong, K.; Lyle, J.; Otto, G.; Peluso, P.; Rank, D.; Baybayan, P.; Bettman, B.; et al. Real-time DNA sequencing from single polymerase molecules. *Science* **2009**, *323*, 133–138. [CrossRef] [PubMed]
2. Rhoads, A.; Au, K.F. PacBio sequencing and its applications. *Genom. Proteom. Bioinform.* **2015**, *13*, 278–289. [CrossRef] [PubMed]
3. Ardui, S.; Ameur, A.; Vermeesch, J.R.; Hestand, M.S. Single molecule real-time (SMRT) sequencing comes of age: Applications and utilities for medical diagnostics. *Nucleic Acids Res.* **2018**, *46*, 2159–2168. [CrossRef] [PubMed]
4. Ameur, A.; Che, H.; Martin, M.; Bunikis, I.; Dahlberg, J.; Hoijer, I.; Haggqvist, S.; Vezzi, F.; Nordlund, J.; Olason, P.; et al. De novo assembly of two Swedish genomes reveals missing segments from the human

GRCh38 reference and improves variant calling of population-scale sequencing data. *Genes* **2018**, *9*, 486. [CrossRef] [PubMed]

5. Suzuki, Y.; Wang, Y.; Au, K.F.; Morishita, S. A statistical method for observing personal diploid methylomes and transcriptomes with single-molecule real-time sequencing. *Genes* **2018**, *9*, 460. [CrossRef]

6. Xiao, C.L.; Zhu, S.; He, M.; Chen, D.; Zhang, Q.; Chen, Y.; Yu, G.; Liu, J.; Xie, S.-Q.; Luo, F.; et al. N^6-methyladenine DNA modification in the human genome. *Mol. Cell* **2018**. [CrossRef]

7. Guiblet, W.M.; Cremona, M.A.; Cechova, M.; Harris, R.S.; Kejnovska, I.; Kejnovsky, E.; Eckert, K.; Chiaromonte, F.; Makova, K.D. Long-read sequencing technology indicates genome-wide effects of non-B DNA on polymerization speed and error rate. *Genome Res.* **2018**, *28*, 1767–1778. [CrossRef]

8. Tilgner, H.; Grubert, F.; Sharon, D.; Snyder, M.P. Defining a personal, allele-specific, and single-molecule long-read transcriptome. *Proc. Natl. Acad. Sci. USA* **2014**, *111*, 9869–9874. [CrossRef]

9. Kuo, R.I.; Tseng, E.; Eory, L.; Paton, I.R.; Archibald, A.L.; Burt, D.W. Normalized long read RNA sequencing in chicken reveals transcriptome complexity similar to human. *BMC Genom.* **2017**, *18*, 323. [CrossRef]

10. Chin, C.S.; Alexander, D.H.; Marks, P.; Klammer, A.A.; Drake, J.; Heiner, C.; Clum, A.; Copeland, A.; Huddleston, J.; Eichler, E.E.; et al. Nonhybrid, finished microbial genome assemblies from long-read SMRT sequencing data. *Nat. Methods* **2013**, *10*, 563–569. [CrossRef]

11. Chin, C.S.; Peluso, P.; Sedlazeck, F.J.; Nattestad, M.; Concepcion, G.T.; Clum, A.; Dunn, C.; O'Malley, R.; Figueroa-Balderas, R.; Morales-Cruz, A.; et al. Phased diploid genome assembly with single-molecule real-time sequencing. *Nat. Methods* **2016**, *13*, 1050–1054. [CrossRef] [PubMed]

12. Sedlazeck, F.J.; Rescheneder, P.; Smolka, M.; Fang, H.; Nattestad, M.; von Haeseler, A.; Schatz, M.C. Accurate detection of complex structural variations using single-molecule sequencing. *Nat. Methods* **2018**, *15*, 461–468. [CrossRef] [PubMed]

13. Wang, M.; Tu, L.; Yuan, D.; Zhu, D.; Shen, C.; Li, J.; Liu, F.; Pei, L.; Wang, P.; Zhao, G.; et al. Reference genome sequences of two cultivated allotetraploid cottons, *Gossypium hirsutum* and *Gossypium barbadense*. *Nat. Genet.* **2018**. [CrossRef]

14. Daccord, N.; Celton, J.M.; Linsmith, G.; Becker, C.; Choisne, N.; Schijlen, E.; van de Geest, H.; Bianco, L.; Micheletti, D.; Velasco, R.; et al. High-quality *de novo* assembly of the apple genome and methylome dynamics of early fruit development. *Nat. Genet.* **2017**, *49*, 1099–1106. [CrossRef] [PubMed]

15. Jiao, Y.; Peluso, P.; Shi, J.; Liang, T.; Stitzer, M.C.; Wang, B.; Campbell, M.S.; Stein, J.C.; Wei, X.; Chin, C.S.; et al. Improved maize reference genome with single-molecule technologies. *Nature* **2017**, *546*, 524–527. [CrossRef] [PubMed]

16. Minio, A.; Lin, J.; Gaut, B.S.; Cantu, D. How single molecule real-time sequencing and haplotype phasing have enabled reference-grade diploid genome assembly of wine grapes. *Front. Plant. Sci.* **2017**, *8*, 826. [CrossRef]

17. Tran, H.T.M.; Ramaraj, T.; Furtado, A.; Lee, L.S.; Henry, R.J. Use of a draft genome of coffee (*Coffea arabica*) to identify SNPs associated with caffeine content. *Plant. Biotechnol. J.* **2018**, *16*, 1756–1766. [CrossRef]

18. Stein, J.C.; Yu, Y.; Copetti, D.; Zwickl, D.J.; Zhang, L.; Zhang, C.; Chougule, K.; Gao, D.; Iwata, A.; Goicoechea, J.L.; et al. Genomes of 13 domesticated and wild rice relatives highlight genetic conservation, turnover and innovation across the genus *Oryza*. *Nat. Genet.* **2018**, *50*, 285–296. [CrossRef]

19. VanBuren, R.; Wai, C.M.; Colle, M.; Wang, J.; Sullivan, S.; Bushakra, J.M.; Liachko, I.; Vining, K.J.; Dossett, M.; Finn, C.E.; et al. A near complete, chromosome-scale assembly of the black raspberry (*Rubus occidentalis*) genome. *Gigascience* **2018**, *7*. [CrossRef]

20. Harkess, A.; Zhou, J.; Xu, C.; Bowers, J.E.; Van der Hulst, R.; Ayyampalayam, S.; Mercati, F.; Riccardi, P.; McKain, M.R.; Kakrana, A.; et al. The asparagus genome sheds light on the origin and evolution of a young Y chromosome. *Nat. Commun.* **2017**, *8*, 1279. [CrossRef]

21. Wang, M.; Wang, P.; Liang, F.; Ye, Z.; Li, J.; Shen, C.; Pei, L.; Wang, F.; Hu, J.; Tu, L.; et al. A global survey of alternative splicing in allopolyploid cotton: Landscape, complexity and regulation. *New Phytol.* **2018**, *217*, 163–178. [CrossRef] [PubMed]

22. Ren, P.; Meng, Y.; Li, B.; Ma, X.; Si, E.; Lai, Y.; Wang, J.; Yao, L.; Yang, K.; Shang, X.; et al. Molecular mechanisms of acclimatization to phosphorus starvation and recovery underlying full-length transcriptome profiling in barley (*Hordeum vulgare* L.). *Front. Plant. Sci.* **2018**, *9*, 500. [CrossRef] [PubMed]

23. Wang, B.; Regulski, M.; Tseng, E.; Olson, A.; Goodwin, S.; McCombie, W.R.; Ware, D. A comparative transcriptional landscape of maize and sorghum obtained by single-molecule sequencing. *Genome Res.* **2018**, *28*, 921–932. [CrossRef] [PubMed]

24. Clavijo, B.J.; Venturini, L.; Schudoma, C.; Accinelli, G.G.; Kaithakottil, G.; Wright, J.; Borrill, P.; Kettleborough, G.; Heavens, D.; Chapman, H.; et al. An improved assembly and annotation of the allohexaploid wheat genome identifies complete families of agronomic genes and provides genomic evidence for chromosomal translocations. *Genome Res.* **2017**, *27*, 885–896. [CrossRef] [PubMed]

25. Dong, L.; Liu, H.; Zhang, J.; Yang, S.; Kong, G.; Chu, J.S.; Chen, N.; Wang, D. Single-molecule real-time transcript sequencing facilitates common wheat genome annotation and grain transcriptome research. *BMC Genom.* **2015**, *16*, 1039. [CrossRef] [PubMed]

26. Fraiture, M.A.; Herman, P.; Papazova, N.; De Loose, M.; Deforce, D.; Ruttink, T.; Roosens, N.H. An integrated strategy combining DNA walking and NGS to detect GMOs. *Food Chem.* **2017**, *232*, 351–358. [CrossRef]

27. Torresen, O.K.; Star, B.; Jentoft, S.; Reinar, W.B.; Grove, H.; Miller, J.R.; Walenz, B.P.; Knight, J.; Ekholm, J.M.; Peluso, P.; et al. An improved genome assembly uncovers prolific tandem repeats in Atlantic cod. *BMC Genom.* **2017**, *18*, 95. [CrossRef]

28. Kalbfleisch, T.S.; Rice, E.S.; DePriest, M.S., Jr.; Walenz, B.P.; Hestand, M.S.; Vermeesch, J.R.; O'Connell, B.L.; Fiddes, I.T.; Vershinina, A.O.; Saremi, N.F.; et al. Improved reference genome for the domestic horse increases assembly contiguity and composition. *Commun. Biol.* **2018**, *1*, 197. [CrossRef]

29. Koren, S.; Rhie, A.; Walenz, B.P.; Dilthey, A.T.; Bickhart, D.M.; Kingan, S.B.; Hiendleder, S.; Williams, J.L.; Smith, T.P.L.; Phillippy, A.M. *De novo* assembly of haplotype-resolved genomes with trio binning. *Nat. Biotechnol.* **2018**. [CrossRef]

30. Bickhart, D.M.; Rosen, B.D.; Koren, S.; Sayre, B.L.; Hastie, A.R.; Chan, S.; Lee, J.; Lam, E.T.; Liachko, I.; Sullivan, S.T.; et al. Single-molecule sequencing and chromatin conformation capture enable *de novo* reference assembly of the domestic goat genome. *Nat. Genet.* **2017**, *49*, 643–650. [CrossRef]

31. Warren, W.C.; Hillier, L.W.; Tomlinson, C.; Minx, P.; Kremitzki, M.; Graves, T.; Markovic, C.; Bouk, N.; Pruitt, K.D.; Thibaud-Nissen, F.; et al. A new chicken genome assembly provides insight into avian genome structure. *G3* **2017**, *7*, 109–117. [CrossRef] [PubMed]

32. Torresen, O.K.; Brieuc, M.S.O.; Solbakken, M.H.; Sorhus, E.; Nederbragt, A.J.; Jakobsen, K.S.; Meier, S.; Edvardsen, R.B.; Jentoft, S. Genomic architecture of haddock (*Melanogrammus aeglefinus*) shows expansions of innate immune genes and short tandem repeats. *BMC Genom.* **2018**, *19*, 240. [CrossRef]

33. Thomas, S.; Underwood, J.G.; Tseng, E.; Holloway, A.K. Bench to basinet CvDC informatics subcommittee. Long-read sequencing of chicken transcripts and identification of new transcript isoforms. *PLoS ONE* **2014**, *9*, e94650. [CrossRef] [PubMed]

34. Dickey, A.M.; Loy, J.D.; Bono, J.L.; Smith, T.P.; Apley, M.D.; Lubbers, B.V.; DeDonder, K.D.; Capik, S.F.; Larson, R.L.; White, B.J.; et al. Large genomic differences between *Moraxella bovoculi* isolates acquired from the eyes of cattle with infectious bovine keratoconjunctivitis versus the deep nasopharynx of asymptomatic cattle. *Vet. Res.* **2016**, *47*, 31. [CrossRef] [PubMed]

35. Zoledowska, S.; Motyka-Pomagruk, A.; Sledz, W.; Mengoni, A.; Lojkowska, E. High genomic variability in the plant pathogenic bacterium *Pectobacterium parmentieri* deciphered from de novo assembled complete genomes. *BMC Genom.* **2018**, *19*, 751. [CrossRef] [PubMed]

36. Aylward, J.; Steenkamp, E.T.; Dreyer, L.L.; Roets, F.; Wingfield, B.D.; Wingfield, M.J. A plant pathology perspective of fungal genome sequencing. *IMA Fungus* **2017**, *8*, 1–15. [CrossRef] [PubMed]

37. A reference standard for genome biology. *Nat. Biotechnol.* **2018**, *36*, 1121. [CrossRef]

38. Kyaschenko, J.; Clemmensen, K.E.; Hagenbo, A.; Karltun, E.; Lindahl, B.D. Shift in fungal communities and associated enzyme activities along an age gradient of managed *Pinus sylvestris* stands. *ISME J.* **2017**, *11*, 863–874. [CrossRef]

39. Heeger, F.; Bourne, E.C.; Baschien, C.; Yurkov, A.; Bunk, B.; Sproer, C.; Overmann, J.; Mazzoni, C.J.; Monaghan, M.T. Long-read DNA metabarcoding of ribosomal RNA in the analysis of fungi from aquatic environments. *Mol. Ecol. Resour.* **2018**, *18*, 1500–1514. [CrossRef]

40. Kronenberg, Z.N.; Fiddes, I.T.; Gordon, D.; Murali, S.; Cantsilieris, S.; Meyerson, O.S.; Underwood, J.G.; Nelson, B.J.; Chaisson, M.J.P.; Dougherty, M.L.; et al. High-resolution comparative analysis of great ape genomes. *Science* **2018**, *360*. [CrossRef]

41. Sutton, J.T.; Helmkampf, M.; Steiner, C.C.; Bellinger, M.R.; Korlach, J.; Hall, R.; Baybayan, P.; Muehling, J.; Gu, J.; Kingan, S.; et al. A high-quality, long-read *de novo* genome assembly to aid conservation of Hawaii's last remaining crow species. *Genes* **2018**, *9*, 393. [CrossRef] [PubMed]

42. Ameur, A.; Kloosterman, W.P.; Hestand, M.S. Single-molecule sequencing: Towards clinical applications. *Trends Biotechnol.* **2019**, *37*, 72–85. [CrossRef] [PubMed]

![genes logo] *genes*

MDPI

Article

Genome Sequencing Illustrates the Genetic Basis of the Pharmacological Properties of *Gloeostereum incarnatum*

Xinxin Wang [1,2,3,†], Jingyu Peng [3,†], Lei Sun [1], Gregory Bonito [3], Jie Wang [4], Weijie Cui [1], Yongping Fu [1,*] and Yu Li [1,*]

[1] Engineering Research Center of Chinese Ministry of Education for Edible and Medicinal Fungi, Jilin Agricultural University, Changchun 130118, China; wangx220@msu.edu (X.W.); sunlei@jlau.edu.cn (L.S.); cuiweijie825@126.com (W.C.)
[2] Department of Plant Protection, Shenyang Agricultural University, Shenyang 110866, China
[3] Department of Plant, Soil, and Microbial Sciences, Michigan State University, East Lansing, MI, USA; pengjin2@msu.edu (J.P.); bonito@msu.edu (G.B.)
[4] Department of Plant Biology and Center for Genomics Enabled Plant Science, Michigan State University, East Lansing, MI, USA; wangjie6@msu.edu
* Correspondence: yongpingfu81@126.com (Y.F.); yuli966@126.com (Y.L.)
† These authors contribute equally to this work.

Received: 17 December 2018; Accepted: 22 February 2019; Published: 1 March 2019

Abstract: *Gloeostereum incarnatum* is a precious edible mushroom that is widely grown in Asia and known for its useful medicinal properties. Here, we present a high-quality genome of *G. incarnatum* using the single-molecule real-time (SMRT) sequencing platform. The *G. incarnatum* genome, which is the first complete genome to be sequenced in the family *Cyphellaceae*, was 38.67 Mbp, with an N50 of 3.5 Mbp, encoding 15,251 proteins. Based on our phylogenetic analysis, the *Cyphellaceae* diverged ~174 million years ago. Several genes and gene clusters associated with lignocellulose degradation, secondary metabolites, and polysaccharide biosynthesis were identified in *G. incarnatum*, and compared with other medicinal mushrooms. In particular, we identified two terpenoid-associated gene clusters, each containing a gene encoding a sesterterpenoid synthase adjacent to a gene encoding a cytochrome P450 enzyme. These clusters might participate in the biosynthesis of incarnal, a known bioactive sesterterpenoid produced by *G. incarnatum*. Through a transcriptomic analysis comparing the *G. incarnatum* mycelium and fruiting body, we also demonstrated that the genes associated with terpenoid biosynthesis were generally upregulated in the mycelium, while those associated with polysaccharide biosynthesis were generally upregulated in the fruiting body. This study provides insights into the genetic basis of the medicinal properties of *G. incarnatum*, laying a framework for future characterization of bioactive proteins and pharmaceutical uses of this fungus.

Keywords: *Gloeostereum incarnatum*; whole genome sequencing; PacBio; secondary metabolite; cytochrome P450 enzyme (CYP); terpenoid

1. Introduction

Mushrooms are an important source of nutrition, and a growing body of evidence has indicated that mushrooms may have medicinal properties and human health benefits [1–3]. *Gloeostereum incarnatum* (family Cyphellaceae) is an edible mushroom, which grows as a saprophyte on broad-leaved trees [4]. *G. incarnatum* is native to China, but is popular in other regions in Asia too, such as Japan and Siberia [4]. Besides its savory taste, *G. incarnatum* is well-known for its medicinal properties. Antioxidant, immunomodulatory, anti-inflammatory, anti-proliferative, and antibacterial properties have been attributed to this mushroom [5–7]. Recent studies have shown that sesquiterpenes and polysaccharides are the main bioactive compounds underlying the beneficial effects of *G. incarnatum* [5,6,8].

With the rapid advancement of sequencing technologies, the number of available fungal genomes has increased [9,10]. However, genomes of medicinal mushrooms remain scarce. Recently, the genomes of a few medicinal mushrooms (e.g., *Ganoderma lucidum*, *Antrodia cinnamomea*, and *Hericium erinaceus*) were released, and proteins putatively associated with the pharmacological properties of these mushrooms were investigated [11–13]. Gene clusters associated with the synthesis of various bioactive secondary metabolites, such as terpenoids and polypeptides, have been identified in many medical mushroom genomes [11,13]. For instance, nine gene clusters associated with the cytochrome P450 (CYP) and triterpenoid pathways were identified in *A. cinnamomea* [11], while four gene clusters associated with terpene and polyketide biosynthesis were identified in *H. erinaceus* [13]. In *G. lucidum*, 24 physical CYP gene clusters, possibly involved in triterpenoid biosynthesis, were identified [12]. Although several bioactive compounds have been identified in *G. incarnatum* [5,6,8,14], the genetic basis of the medicinal benefits of this mushroom are largely unknown.

In this study, we used the Pacific Biosciences (PacBio) long-read sequencing platform [15] to perform the de novo assembly of the *G. incarnatum* genome. This is the first genome to be sequenced in the Cyphellaceae family. We also compared the transcriptome profiles of the mycelium and the fruiting body, the two major developmental stages of *G. incarnatum*. The sequenced genome of *G. incarnatum* presented herein is, to our knowledge, one of the most comprehensive assembled genomes of an edible mushroom. In this study, we aimed to (1) present a high-quality reference genome for *G. incarnatum*, which can be used for future analyses of genome function and genetic variation and (2) identify relevant functional genes, gene clusters, and signaling pathways associated with the saprophytic lifestyle and pharmaceutical properties of *G. incarnatum*. We specifically focused on terpene biosynthesis, cytochrome P450 enzyme biosynthesis, and polysaccharide production. Our study provides a valuable genomic and transcriptomic resource for future studies of the genetic basis of the medicinal properties of *G. incarnatum*. Such studies would represent a first step towards realizing the full potential of *G. incarnatum* as a source of pharmacologically active compounds on an industrial scale.

2. Materials and Methods

2.1. Fungal Material, Sequencing, and Genome Assembly

We isolated protoplast-derived monokaryons from the dikaryotic strain of the *G. incarnatum* commercial strain CCMJ2665. The monokaryons were obtained as described previously [16], except that the dikaryotic mycelia were incubated for 240 min at 30 °C in lywallzyme lysing enzyme. The single-nucleated genomic DNA of the *G. incarnatum* monokaryon strain was then used for genome sequencing and annotation. Genomic DNA was extracted using NuClean Plant Genomic DNA Kits (CWBIO, Beijing, China). The genome of *G. incarnatum* was sequenced on a PacBio Sequel long-read sequencing platform with a library insert size of 20 kb, at the Engineering Research Center of the Chinese Ministry of Education for Edible and Medicinal Fungi, Jilin Agricultural University (Changchun, China). Raw data were assembled with SMARTdenovo (https://github.com/ruanjue/smartdenovo). The completeness of the genome assembly was evaluated using Core Eukaryotic Genes Mapping Approach (CEGMA) [17] and Benchmarking Universal Single-Copy Orthologs (BUSCO; [18]). The whole-genome sequence of *G. incarnatum* has been deposited in GenBank (in submission). The genome reported in this study has been deposited in GenBank under the accession RZIO00000000.

2.2. Genome Annotation

Three different strategies were used to predict genes in the *G. incarnatum* genome: Sequence homologies with four representative mushrooms; ab initio with Augustus [19], Genescan [20], GlimmerHMM [21], and SNAP [22]; and combining extrinsic and ab initio approaches with

GLEAN (http://sourceforge.net/projects/glean-gene). GLEAN gene prediction results were used for subsequent analyses. Protein-coding genes were annotated by GLEAN using both ab initio and evidence-based methods [23]. Predicted genes were functionally annotated against several databases—National Center for Biotechnology Information (NCBI) non-redundant (nr), Swiss-Prot, and InterPro—using BLASTP searches (e-value $\leq 1 \times 10^{-5}$). Gene annotations were refined using the following databases: Gene Ontology (GO) [24], Clusters of Orthologous Groups (KOG) [25], and Kyoto Encyclopedia of Genes and Genomes (KEGG) [26]. Transposon sequences were identified by aligning the assembled genome to the Repbase database [27] with RepeatMasker (version 3.3.0; http://www.repeatmasker.org/; [28]) and RepeatProteinMasker [22]. Tandem repeat sequences (TRF) were predicted with Tandem Repeat Finder [29]. Ribosomal RNA (rRNA) sequences were identified, based on sequence homology and also through use of de novo prediction strategies with rRNAmmer [30]. Transfer RNA (tRNA) genes were identified using tRNAscan-SE [31]. Non-coding RNAs, such as small nuclear RNA (snRNAs) and microRNAs (miRNAs), were predicted with Rfam [32].

2.3. Evolutionary Analysis and Phylogeny

The phylogenetic analysis was performed using single-copy genes shared across *G. incarnatum* and another nine fungal species (*Omphalotus olearius, Gymnopus luxurians, Laccaria bicolor, Coprinopsis cinerea, Armillaria ostoyae, Lentinula edodes, Schizophyllum commune, Serpula lacrymans* and *Coniophora puteana*). The "all against all" BLASTP searches were performed with a cutoff e-value of 1×10^{-7} for proteins from all species. The alignments of gene pairs were conjoined by solar. Only gene pairs with an alignment ratio (aligned region by total length) of more than 30% in both homologous genes were kept for the following gene family construction. Gene families were clustered using a sparse graph of gene relationships using the hierarchical clustering algorithm hcluster_sg 0.5.1 package. Finally, we identified single-copy genes which had only one homolog per taxon, and those genes were used to construct the phylogenetic tree. The protein sequences of these single-copy genes were aligned using MUSCLE [33] and the protein alignments were transformed into codon alignments with PAL2NAL. Gblocks was used to refine each codon alignment, and all refined alignments were concatenated to a super codon alignment. RAxML software (version 7.2.3) [34] was used to construct the phylogenetic tree using the maximum likelihood (ML) algorithm. The best-scoring ML tree was inferred using the rapid bootstrap analysis after 1000 runs. The divergence times among species were estimated using the mcmctree module in PAML [35] with the calibration time of *Serpula lacrymans* and *Coniophora puteana* according to Floudas et al. (2012).

2.4. Carbohydrate-Active Enzyme (CAZyme) Family Classification

The CAZymes in the *G. incarnatum* genome were identified by mapping the annotated protein sequences to the CAZy database (http://www.cazy.org/) [36] using BLASTP (cut-off e-value $\leq 1 \times 10^{-5}$, identity $\geq 40\%$ and coverage $\geq 40\%$). The recovered CAZymes were classified as glycoside hydrolases (GHs), auxiliary activities (AAs), carbohydrate-binding modules (CBMs), glycosyl transferases (GTs), polysaccharide lyases (PLs), and carbohydrate esterases (CEs).

2.5. Cytochrome P450 (CYP) Predictions

CYP proteins were predicted by aligning the gene models to the fungal P450 database (http://p450.riceblast.snu.ac.kr/index.php?a=view;) with BLASTP (e-value $\leq 1 \times 10^{-5}$, matrix = BLOSUM62). CYP proteins were assigned to protein families based on Nelson's nomenclature [37]. For protein sequences that aligned with multiple families, the top hit was chosen.

2.6. Secondary Metabolite Annotations

Secondary metabolite gene clusters were predicted with fungal AntiSMASH 3.0 (https://fungismash.secondarymetabolites.org/) [38], with the default parameter values.

2.7. RNA Sequencing of the Two Major Developmental Stages

Samples of the two major fungal developmental stages (mycelium and fruiting body) from the *G. incarnatum* strain CCMJ2665 were provided by the mushroom section of the Engineering Research Center of the Chinese Ministry of Education for Edible and Medicinal Fungi, Jilin Agricultural University (Changchun, China). RNA extraction and quality control were performed following the processes of Fu et al. [16]. cDNA libraries were constructed, and 150 paired-end sequencing was performed on an Illumina HiSeq 4000 platform at Novogene Co., LTD (Beijing, China). Sequencing data have been deposited in the NCBI SRA (accession no. PRJNA510218).

Raw data were filtered to remove adapter sequences and low-quality reads for downstream analyses. The trimmed reads were mapped to the *G. incarnatum* genome using TopHat v2.0.12 [39]. The number of reads mapped to each gene was counted using HTSeq v0.6.1 [40]. Fragments per kilobase of transcript per million mapped reads (FPKM) values were used to calculate gene expression. Genes differentially expressed between developmental stages were identified using the DESeq package (1.18.0) [41] in R with adjusted *p*-value set to <0.05.

3. Results and Discussion

3.1. Genome Sequencing and Assembly

A high-quality reference genome for *G. incarnatum* was generated from a protoplast monokaryon isolated from the dikaryotic strain of a commercial *G. incarnatum* cultivar (CCMJ2665; see Table 1). The genomic DNA of *G. incarnatum* was sequenced on PacBio SMRT Sequel platform generating ~94× coverage of 3,642 Mbp of clean data, as shown in Table S1. Compared to other edible and medicinal mushrooms, the assembled genome of *G. incarnatum* (38.7 Mbp), as shown in Figure 1, was of an intermediate size; the *Wolfiporia cocos* genome was the largest (50.5 Mbp); and the *Agaricus bisporus* var. *bisporus* genome was the smallest (30.2 Mbp), as shown in Table 1 [11–13,42–48]. *G. incarnatum* had a guanine-cytosine (GC) content of 49%; GC content in the other mushroom genomes examined was 45.3–55.9% (Table 1). The genome of *G. incarnatum* was one of the most complete assembled genomes across all representative edible and medicinal mushrooms examined, consisting of 20 scaffolds with an N50 of 3.5 Mbp (Table 1; Figure 1). The completeness of the *G. incarnatum* genome assembly was analyzed with the CEGMA [17] and the single-copy orthologs test using Fungi BUSCOs [18]. The CEGMA analysis indicated that 96.8% of the core eukaryotic genes were mapped to the *G. incarnatum* genome. The BUSCO analysis suggested that the annotation set was well completed, with 93.1% complete BUSCOs and 4.5% missing BUSCOs. Thus, our results indicate that the *G. incarnatum* genome assembly is high quality.

Table 1. Comparison of genome assembly among representative edible mushrooms.

Organism	Accession	Genome Size (Mbp)	Genome	Scaffold	N50 (Kbp)	GC Content (%)	Protein-Coding Genes	Sequencing Method
Gloeostereum incarnatum		38.7	94×	20	3500	49.0	15,251	PacBio Sequel
Lentinula edodes	LSDU00000000	46.1	60×	31	3663	45.3	13,426	PacBio RSII; Illumina HiSeq 2500
Agrocybe aegerita	PRJEB21917	44.8	253×	122	768	49.2	14,113	PacBio RSII; Illumina HiSeq 2500
Hericium erinaceus	PRJN361338	39.4	200×	519	538	53.1	9895	Illumina MiSeq; Hiseq 2500
Antrodia cinnamomea	JNBV00000000	32.2	878×	360	1035	50.6	9254	Roche 454; Illumina GAIIx
Ganoderma lucidum	AGAX00000000	43.3	440×	82	1388	55.9	16,113	Roche 454; Illumina GAII
Wolfiporia cocos	AEHD00000000	50.5	40×	348	2539	52.2	12,212	Sanger; Roche 454
Inonotus baumii	LNZH00000000	31.6	186×	217	267	47.6	8455	Illumina HiSeq
Agaricus bisporus var. *bisporus*	AEOK00000000	30.2	8.3×	29	2300	46.6	10,438	Sanger
Lignosus rhinocerotis	AXZM00000000	34.3	180×	1338	90	53.7	10,742	Illumina Hiseq 2000
Sparassis latifolia	LWKX00000000	48.1	601×	472	641	51.4	12,471	Illumina HiSeq 2500
Flammulina velutipes	BDAN00000000	35.3	132×	5130	150	49.6	13,843	Illumina HiSeq 2500

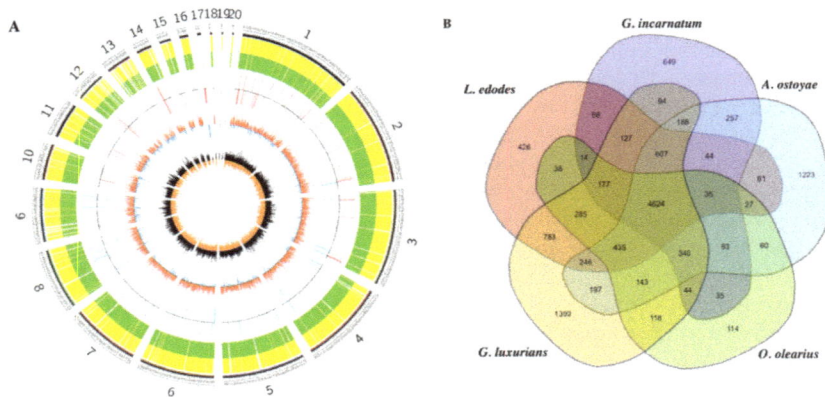

Figure 1. The *Gloeostereum incarnatum* genome and comparative genomics analysis. (**A**) The *G. incarnatum* genome. Outside to inside of concentric circles show assembly scaffold number, gene density, non-coding RNA (ncRNA), GC count and GC skew. (**B**) Unique and homologous gene families. The number of unique and shared gene families is shown in each of the diagram components and the total number of gene families for each fungus is given in parentheses.

3.2. Gene and Repeat Sequence Prediction and Annotation

To most accurately predict the protein-coding genes in the *G. incarnatum* genome, we used a homology-based prediction strategy (against four representative mushroom genomes) combined with de novo gene prediction approaches. We predicted 15,251 protein-coding genes, accounting for 57.46% of the assembled *G. incarnatum* genome (Table S2). The predicted protein-coding genes had an average length of 1456.86 bp and contained 4.38 exons (each with an average length of 264.46 bp). The protein-coding genes were functionally annotated against several databases: NCBI nr, Swiss-Prot, InterPro, GO, COG, and KEGG. Of the 15,251 protein-coding genes predicted, 72.62% had homologs in one or more of the databases searched (Table S2).

We identified ~5.9 Mbp of repeat sequences in the *G. incarnatum* genome. Of these repeat sequences, 0.49% were predicted to be tandem repeats and 14.46% to be transposons (TEs) (Figure 1; Table S3). Most of the predicted TEs were long terminal repeats (LTRs), representing 13.78% of the genome (Table S3). Of the non-coding RNA species we identified in the *G. incarnatum* genome, 161 were tRNAs and 44 were rRNAs (Table S4). Nine of the identified tRNAs were possible pseudogenes, and the remaining 152 anti-codon tRNAs corresponded to the 20 common amino acids (Table S4). We also predicted 18 miRNAs and 18 snRNAs; the snRNAs comprised 15 spliceosomal RNAs and three C/D box small nucleolar RNAs (Table S4).

KOG functionally classified 4243 (27.82%) of the predicted proteins. Of these, 499 genes were associated with "amino acid transport and metabolism", 476 genes were associated with "carbohydrate transport and metabolism", and 337 with "secondary metabolite biosynthesis, transport, and catabolism". This suggests that several *G. incarnatum* proteins are involved in nutrient absorption, transformation, and the synthesis of secondary metabolites. Similarly, KEGG classification indicated that both "amino acid metabolism" and "carbohydrate metabolism" were enriched in *G. incarnatum* genes (603 and 654 genes, respectively). KEGG analysis indicated that another 161 proteins were assigned to "biosynthesis of other secondary metabolites", and 59 proteins were associated with the "metabolism of terpenoids and polyketides". As the medicinal properties of edible mushrooms are closely related to the biosynthesis of secondary metabolites [49], these compounds were the focus of the remainder of our study.

3.3. Comparative Genomics and Evolutionary Analysis

With the exception of the *G. incarnatum* genome assembled in this study, no complete genomes are available for other fungi in Cyphellaceae. Thus, our evolutionary analysis compared whole genome sequences of seven representative species of the Agaricales: *O. olearius*, *G. luxurians*, *L. bicolor*, *C. cinerea*, *A. ostoyae*, *L. edodes*, and *S. commune*. We also included genomes of two additional agaricomycetid species having fossil calibrations—*S. lacrymans* and *C. puteana* [50]. We found that the *G. incarnatum* genome includes 7384 gene families, with 10,075 (66.1%) genes having homologs in at least one of the other nine fungal species (Figure 1, Table S5). Interestingly, 5,176 (33.9%) unclustered genes and 469 unique gene families (containing 1369 genes) were *G. incarnatum* specific (Table S5). These *G. incarnatum*-specific genes were associated with diverse biological processes, including steroid biosynthesis, terpenoid backbone biosynthesis, and polysaccharide biosynthesis.

We then constructed an ML phylogeny for *G. incarnatum* and the nine additional fungal species, based on 1822 shared single-copy orthologous genes (Figure 2). These data indicate that *G. incarnatum* is phylogenetically closer to *A. ostoyae*, diverging ~174 million years ago (Figure 2). We also identified 325 significantly expanded gene families in the *G. incarnatum* genome ($p \leq 0.01$) (Figure 2); these families were primarily associated with carbohydrate metabolism (starch/sucrose metabolism and glycolysis/gluconeogenesis), amino acid and lipid metabolism, and genetic and environmental information processing. However, caution is warranted when interpreting species' evolutionary time estimates and gene family expansions and contractions based on genomes generated with different sequencing platforms, assembly methods, and selection of comparative analysis groups. Nevertheless, our analysis provides new insights into the phylogeny of *G. incarnatum* and other mushroom species based on whole-genome data.

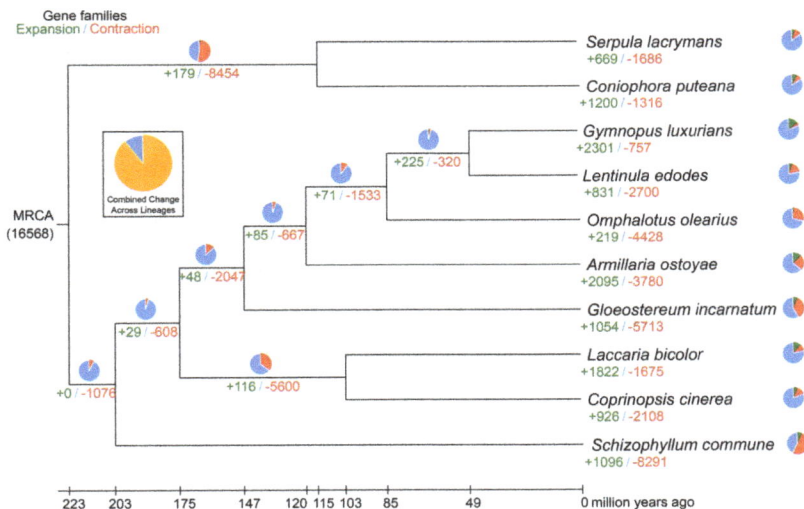

Figure 2. The *Gloeostereum incarnatum* genome evolutionary analysis. The number of expanded (green) and contracted (red) gene families is shown at each branch. The estimated divergence time (MYA: million years ago) is shown at the bottom. MRCA: most recent common ancestor.

3.4. The Decomposition of Wood by CAZymes

To further classify the proteins associated with lignin digestion during carbohydrate metabolism, we mapped the protein sequences of *G. incarnatum* to the CAZy database [36]. We identified 311 non-overlapping CAZymes in six families in *G. incarnatum* (Table S6); the CAZymes in *G. incarnatum* were more diverse and abundant than those of brown rot fungi [51]. The *G. incarnatum* CAZymes consisted of 164 GHs, 66 proteins with AAs, 42 CBMs, 41 GTs, 18 PLs, and 10 CEs (Figure 3).

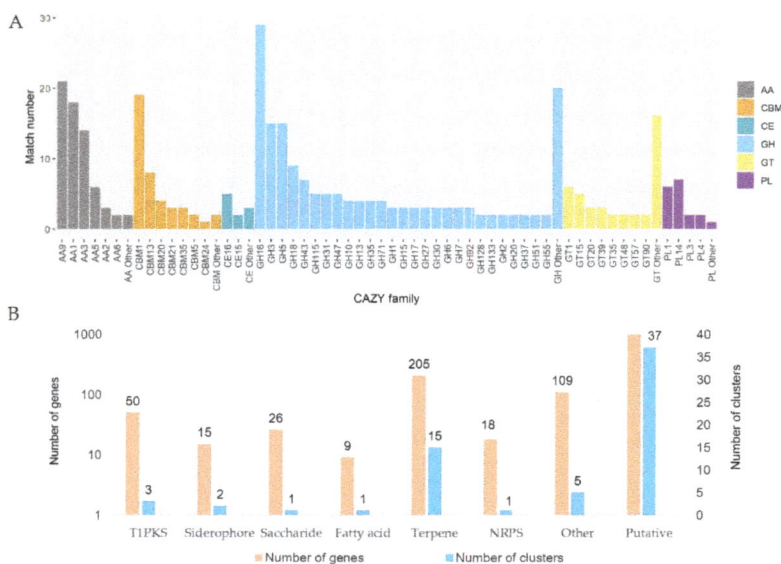

Figure 3. (**A**) Annotation of carbohydrate-related genes in the *G. incarnatum* genome; (**B**) secondary metabolite-related gene clusters in the *G. incarnatum* genome. T1PK: Type I polyketide synthases; NRPS: nonribosomal peptide synthetase.

As the GHs include many cellulase families (such as GH16, GH5, GH3, GH6, and GH7) [36], the remarkably higher number of GHs in *G. incarnatum* was not unexpected. As a saprotrophic mushroom, *G. incarnatum* is likely to require many GHs to decompose cellulose from its woody hosts. AAs were the next most abundant family of CAZymes in *G. incarnatum*; AAs identified in this species included 21 AA9, 18 AA1, and 14 AA3 enzyme families. These three AA families are also the most abundant AAs in other fungi [10–13]. However, only three AA2 family proteins, the lignin-modifying fungal peroxidases (PODs), were identified in *G. incarnatum*. PODs are the primary lignin decomposers in the model white rot fungus *Phanerochaete chrysosporium* and other fungal species [52]. As *G. incarnatum* is restricted to elm tree hosts, the few AA2s identified in this fungus may be sufficient to decompose lignins produced by elm. We also identified 21 genes encoding enzymes for pectin digestion in *G. incarnatum*. Thus, the wood-decaying mushroom, *G. incarnatum*, may utilize complex strategies to decompose plant cell walls.

3.5. Secondary Metabolites and Terpene Pathway

The pharmacological properties of medicinal mushrooms are largely conferred by secondary metabolites; these metabolites have received intense research attention [1,49,53]. Here, we used antiSMASH to search for gene clusters encoding secondary metabolites in *G. incarnatum* [38]. We identified 65 gene clusters: one saccharide, 15 terpene synthases (TSs), one fatty acid, three polyketide synthases (PKSs), two siderophores, one non-ribosomal peptide-synthetase (NRPS), and 37 putative gene clusters of unknown type (Figure 3).

Terpenoid biosynthesis is of particular interest as terpenoids are important pharmacologically active compounds in *G. incarnatum* [14,54]. *G. incarnatum* contains two unique sesquiterpene compounds, gloeosteretriol and incarnal [14,54]. Both compounds demonstrate antibacterial activity against the Gram-positive bacteria *Staphylococcus aureus* and *Bacillus subtilis*, but not against any of the Gram-negative bacteria tested to date [14,54]. Incarnal extracted from *G. incarnatum* and another fungus in Cyphellaceae (*Chondrostereum* sp.) also shows potent cytotoxicity against several cancer cell lines [8,55]. To further investigate the biosynthesis of terpenoids in *G. incarnatum*, we mapped 35 proteins to 17 enzymes in the "terpenoid backbone biosynthesis" (KEGG: ko00900) pathway, and 10 proteins to four enzymes in the "sesquiterpenoid and triterpenoid biosynthesis" (KEGG: ko00909) pathway (Figure 4). The pathway mapping results suggested that the *G. incarnatum* terpenoids are likely to be synthesized through the mevalonate (MVA) pathway, not the 2-C-methyl-D-erythritol 4-phosphate/1-deoxy-D-xylulose 5-phosphate (MEP/DOXP) pathway. This is in line with the results for other mushrooms, such as *G. lucidum* and *H. erinaceus* [11,12].

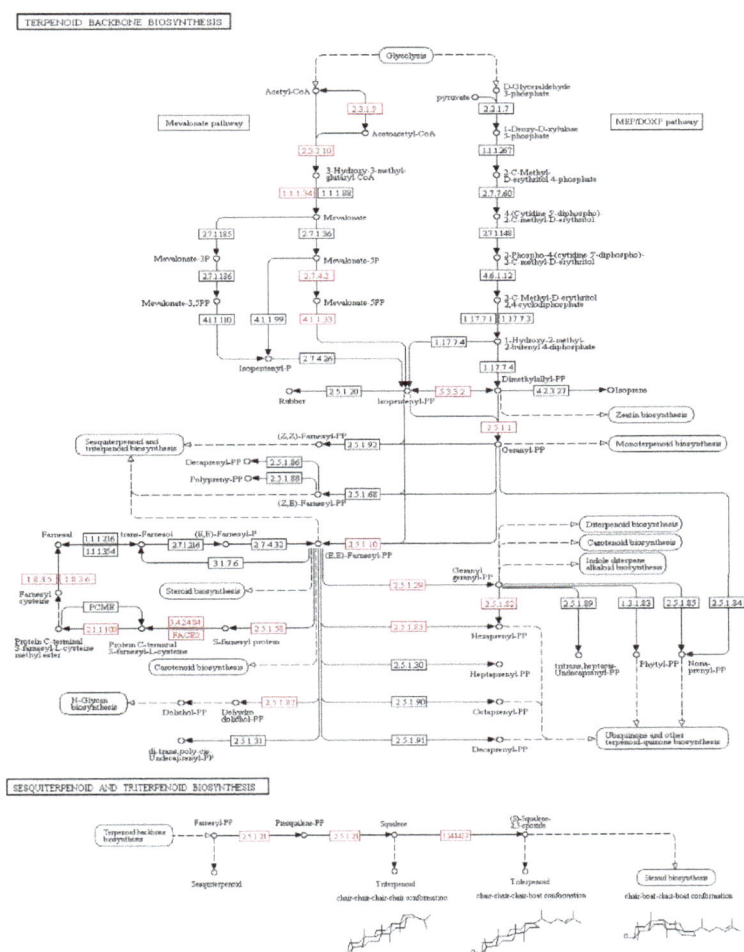

Figure 4. "Terpenoid backbone biosynthesis" (KEGG: ko00900) and "sesquiterpenoid and triterpenoid biosynthesis" (KEGG: ko00909) pathways of *G. incarnatum*. Red boxes indicate the presence of the enzymes, whereas white boxes indicate enzyme is not present.

Sesquiterpenoids are synthesized from farnesyl diphosphate (FPP) by various sesquiterpene synthases [56]. We located five genes encoding sesquiterpene synthases in *G. incarnatum*—three encoding trichodiene synthase (EC 4.2.3.6) and two encoding aristolochene synthase (EC 4.2.3.9; Figure 4). Interestingly, trichodiene is a precursor for the biosynthesis of the mycotoxin nivalenol, which is widely found in *Fusarium* species, and the biosynthesis of aristolochene, which is a precursor for the PR toxin found in *Penicillium* species [57,58]. To the best of our knowledge, neither of these compounds are produced by *G. incarnatum*. It would be interesting to know if trichodiene or aristolochene was a substrate for the synthesis of gloeosteretriol or incarnal in *G. incarnatum*. Based on structural similarity, incarnal might potentially be synthesized from trichodiene in conjunction with certain cyclization, bond-shift rearrangement, oxidation, and hydroxylation reactions. Further experiments are thus necessary to confirm the production of trichodiene and aristolochene, as well as their association with the biosynthesis of gloeosteretriol or incarnal, in *G. incarnatum*.

Regarding triterpenoid, two farnesyltransferases (EC 2.5.1.21), three squalene monooxygenases (EC 1.14.14.17), and one lanosterol synthase (EC 5.4.99.7) were encoded in the *G. incarnatum* genome. This suggested that *G. incarnatum* synthesizes squalene, (S)-2,3-epoxysqualene, and lanosterol, all of which are intermediates in the synthesis of triterpenoid and sterol [59]. Notably, lanosterane-type triterpenoids are produced by several medical mushrooms, including species of *Ganoderma*, *Innonotus*, and *Antrodia* (reviewed in [60]), although the relevant biosynthesis pathways are unknown. The existence of these triterpenoid-related proteins in *G. incarnatum* suggests that this species may produce previously uncharacterized triterpenoids.

3.6. The CYP Family

Although the pathway for terpenoid backbone biosynthesis in fungi is relatively well studied [61,62], the steps following terpenoid cyclization are largely unknown. The structural diversity of terpenoids depends on post-modification of many specific chemical groups. These modifications involve a series of hydroxylation, reduction, oxidation, and acylation reactions, largely mediated by CYPs (cytochrome P450s) [63–65]. In fungi, CYPs are especially important for xenobiotic degradation and the biosynthesis of several secondary metabolites, including terpenoids and polyketides [63]. Based on a comparative search of the Fungal Cytochrome P450 Database [66], 145 CYP proteins were identified in *G. incarnatum*. These proteins were classified into 57 families following Nelson's nomenclature [37]. The family CYP5144 included the greatest number of *G. incarnatum* CYPs (16); CYP5144 also included the most CYPs in another medicinal mushroom, *Lignosus rhinocerotis* [46] (Table 2). It is thus likely that CYP5144 family proteins play key roles in the biosynthesis of terpenoids in *G. incarnatum*.

As previously noted, the *G. incarnatum* genome encoded two sesterterpenoid synthases—aristolochene synthase and trichodiene synthase. Interestingly, the trichodiene synthase genes (GI_10004653, GI_10004654, and GI_10004694), but not the aristolochene synthase gene, (GI_10003231) were identified in gene clusters containing several CYPs (i.e., CYP65X, CYP530A, and CYP617B; Figure 5). Based on the logic in the Fungal Cytochrome P450 Database (FDPD) pipeline [66], CYP530A and CYP617B were assigned to the families CYP512 and CYP5144, respectively. Both of these families may be involved in the biosynthesis of bioactive terpenoids in *G. lucidum* and *L. rhinocerotis* [12]. These results further support our hypothesis that incarnal, the bioactive sesterterpenoid produced by *G. incarnatum*, might be synthesized from trichodiene, mediated by CYPs. The second-largest family of CYPs identified in *G. incarnatum* was CYP620 (Table 2), which is a relatively rare family in other medicinal mushrooms (absent in *G. lucidum*, one in *A. cinnamomea*, and three in *L. rhinocerotis*) [11,12,46]. This indicates that CYP distributions and functions vary among medicinal mushrooms. The exact roles of the identified CYPs in terpenoid post-modification or other biological functions remain to be experimentally validated.

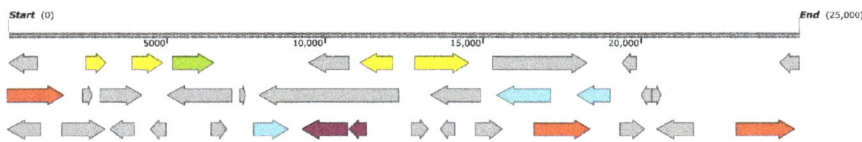

Figure 5. Genetic structures of sesterterpenoid synthase genes and their neighboring genes. Each gene is represented by an arrow. The aristolochene synthase gene (GI_10003231) is indicated by green color; the trichodiene synthase genes (GI_10004653, GI_10004654 and GI_10004694) are indicated by light blue color; the cytochrome P450 (CYP) genes are indicated by red color; choline dehydrogenase genes are indicated by yellow color; The Sec1-like protein genes are indicated by purple color.

Table 2. Summary of the CYP genes in the *G. incarnatum* genome.

Family	Subfamily	Corresponding Gene Number	Total Gene Number	Family	Subfamily	Corresponding Gene Number	Total Gene Number
CYP5144	C,F	15,1	16	CYP675	A	3	3
CYP620	A,B,E,H	1,1,4,2	8	CYP682	B	3	3
CYP5015	C	6	6	CYP504	A	3	3
CYP5014	F,H	2,3	5	CYP51	F	3	3
CYP5068	B	5	5	CYP55	A	3	3
CYP5080	B,D	3,2	5	CYP65	J,X	1,1	2
CYP5093	A	5	5	CYP5070	A	2	2
CYP505	C,D	3,1	4	CYP5074	A	2	2
CYP535	A	4	4	CYP5078	A	2	2
CYP536	A	4	4	CYP5081	A	2	2
CYP617	A,B	1,2	3	CYP5125	A	2	2
CYP5037	B	3	3	CYP540	B	2	2
CYP5110	A	3	3	CYP630	B	2	2
CYP530	A	3	3	Others	-	-	30

3.7. Polysaccharide Biosynthesis

Composition of *G. incarnatum* polysaccharides also had immunomodulatory and immuno-enhancing effects in a mice model [6]. Some of the most potent immunomodulatory polysaccharides produced by medical mushrooms are water soluble 1,3-β- and 1,6-β-glucans [67]. In *G. incarnatum*, we identified four 1,3-β-glucan synthases (K00706 and K01180), three UTP–glucose-1-phosphate uridylyltransferases (K00963), 12 GTPase-activating-associated proteins (K12492, K12493, K19838, K19844, K19845, K14319, K17265, K18470, K20315, and K19839), two hexokinases (K00844), and two phosphoglucomutases (K01835) (Table 3). We also identified 15 β-glucan biosynthesis-associated proteins (PF03935; Table 3); β-glucan biosynthesis-associated proteins were shown to be involved in the biosynthesis of 1,6-β-glucans in *Saccharomyces cerevisiae* [68]. The polysaccharide biosynthesis-related proteins identified in *G. incarnatum* are summarized in Table S5. Compared with five other species of medicinal mushrooms (*Auricularia heimuer* [69], *A. cinnamomea* [11], *Sparassis latifolia* [47], *L. rhinocerotis* [46], and *G. lucidum* [12]), *G. incarnatum* produced more 1,3-β-glucan synthases, GTPase-activating-associated proteins, and β-glucan biosynthesis-associated proteins, as well as similar numbers of UTP–glucose-1-phosphate uridylyltransferases, hexokinases, and phosphoglucomutases (Table S7). This suggests that *G. incarnatum* might produce more 1,3-β- and 1,6-β-glucans. In-parallel quantifications of 1,3-β- and 1,6-β-glucan production among these medicinal mushrooms during different growth phases should be performed and compared.

Table 3. Summary of the polysaccharide biosynthesis-related proteins in *G. incarnatum*.

Enzyme Family	KO Term	EC Number	Gene Number	Gene Name
1,3-β-glucan synthase	K01180	EC:3.2.1.6	1	GI_10004256
	K00706	EC:2.4.1.34	3	GI_10014134, GI_10014600, GI_10010064
UTP–glucose-1-phosphate uridylyltransferase	K00963	EC:2.7.7.9	3	GI_10009949, GI_10009950, GI_10009951
Hexokinase	K00844	EC:2.7.1.1	2	GI_10010509, GI_10009252
Phosphoglucomutase	K01835	EC:5.4.2.2	2	GI_10003989, GI_10014463
GTPase-activating-associated protein	K12492	-	1	GI_10009154
	K19838	-	1	GI_10009280
	K12493	-	1	GI_10004440
	K14319	-	1	GI_10004658
	K19845	-	2	GI_10004984, GI_10007354
	K19839	-	3	GI_10003462, GI_10010380, GI_10012746
	K19844	-	2	GI_10014590, GI_10000357
	K18470	-	1	GI_10014667

3.8. Transcriptomic Analysis

As the expression levels of target genes encoding pharmacologically relevant proteins in *G. incarnatum* might differ across developmental stages, we profiled the transcriptomes of two major developmental stages of *G. incarnatum*—the mycelium and the fruiting body. We generated 70,634,952 raw reads from the cDNA libraries of the two stages. After data filtering and trimming, 69,716,944 high-quality clean reads remained. Of these clean reads, 92% were successfully mapped to the *G. incarnatum* genome. Across both stages, 11,015 genes were expressed, with 944 genes expressed only in the mycelium, and 718 genes only expressed in the fruiting body (Figure 6). We identified 3524 differentially expressed genes (DEGs) in the fruiting body as compared to the mycelium (Figure 6). Of these 1822 were significantly upregulated in the fruiting body as compared to the mycelium, and 1702 were significantly downregulated (Figure 6).

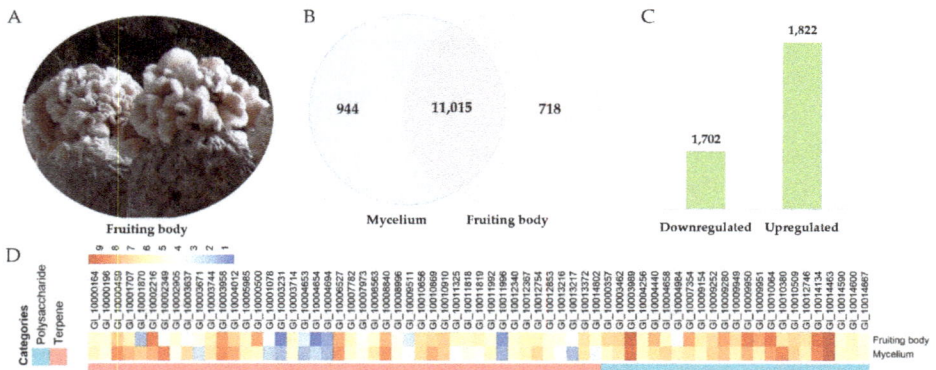

Figure 6. Comparative transcriptome profiling of the mycelium and the fruiting body of *G. incarnatum*: (**A**) Fruiting bodies of *G. incarnatum*; (**B**) Venn diagram of the genes expressed in the mycelium and/or the fruiting body; (**C**) number of genes being significantly downregulated or upregulated in the fruiting body compared with the mycelium; (**D**) heatmap of the genes associated with the biosynthesis of polysaccharides and terpenes.

The gene expression patterns in the mycelium and fruiting body of *G. incarnatum* varied depending on the type of secondary metabolite encoded. For example, 17 of 45 terpenoid biosynthesis-related genes (38%) were differentially expressed between the mycelium and the fruiting body (Figure 6, Table S8). Of these, 65% were upregulated in the mycelium as compared to the fruiting body (Figure 6, Table S8), indicating that the biosynthesis of terpenoid compounds might be greater in the mycelium of *G. incarnatum*. In contrast, 10 of the 23 genes associated with polysaccharide biosynthesis (43%) were differentially expressed between the mycelium and the fruiting body, with 70% of these being significantly upregulated in the fruiting body as compared to the mycelium (Figure 6). This indicates that the fruiting body of *G. incarnatum* might be a richer source of polysaccharides. These findings were consistent with those for terpenoid- and polysaccharide-related genes in *H. erinaceus* [13]. Therefore, different secondary metabolites might be more enriched at different fungal development stages. Due to the complexity of secondary metabolite biosynthesis, further studies should investigate the molecular mechanisms underlying the secondary metabolism of *G. incarnatum*.

4. Conclusions

In this study, we presented the first whole-genome sequence of *G. incarnatum*, which is the first sequenced genome for a fungus belonging to Cyphellaceae. The *G. incarnatum* genome is one of the most completely assembled edible mushroom genomes available to date, consisting of 20 scaffolds with an N50 of 3.5 Mbp. The remarkably higher number of GHs and AAs in *G. incarnatum* may contribute to the active decomposition of lignin and cellulose of its woody hosts. We identified 65 gene clusters involved in the biosynthesis of secondary metabolites in the *G. incarnatum* genome. We also investigated the functions of the proteins involved in terpenoid biosynthesis; terpenoids are one of the main types of pharmacologically active compounds produced by *G. incarnatum*. We found two sesquiterpenoid synthase genes, one encoding aristolochene synthase and the other encoding trichodiene synthase, in gene clusters enriched with CYP genes. This suggested that CYPs play an active role in the post-modification of aristolochene and trichodiene sesquiterpenoids. We also predicted 38 proteins involved in polysaccharides biosynthesis, another main class of bioactive compounds in *G. incarnatum*. Genes involved in terpenoid biosynthesis were generally upregulated in mycelium, while the polysaccharide biosynthesis-related genes were upregulated in the fruiting body. These results provide a foundation for future studies of the genetic basis underlying the medicinal properties of *G. incarnatum*.

Supplementary Materials: Supplementary materials can be found at http://www.mdpi.com/2073-4425/10/3/188/s1.

Author Contributions: Conceptualization and supervision, Y.L. and Y.F.; sample preparation, W.C.; formal analysis and writing, J.P. and X.W.; software, J.P., L.S. and J.W.; critical review, G.B.

Funding: This research was funded by the Special Fund for Agro-Scientific Research in the Public Interest (No. 201503137); the Program of Creation and Utilization of Germplasm of Mushroom Crop of "111" Project (No. D17014); National-Level International Joint Research Center (2017B01011).

Conflicts of Interest: The authors declare no conflict of interest.

Abbreviations

SMRT	Single-Molecule, Real-Time
KEGG	Kyoto Encyclopedia of Genes and Genomes
CYP	cytochrome P450
CAZymes	carbohydrate-active enzymes
DEG	differentially expressed genes
FPKM	fragments per kilobase of transcript per million mapped reads

References

1. Wasser, S.P. Medicinal mushroom science: Current perspectives, advances, evidences, and challenges. *Biomed. J.* **2014**, *37*, 345–356. [CrossRef] [PubMed]
2. Song, C.; Liu, Y.; Song, A.; Dong, G.; Zhao, H.; Sun, W.; Ramakrishnan, S.; Wang, Y.; Wang, S.; Li, T.; et al. The *Chrysanthemum nankingense* genome provides insights into the evolution and diversification of chrysanthemum flowers and medicinal traits. *Mol. Plant* **2018**, *11*, 1482–1491. [CrossRef] [PubMed]

3. Guggenheim, A.G.; Wright, K.M.; Zwickey, H.L. Immune modulation from five major mushrooms: Application to integrative oncology. *Integr. Med.* **2014**, *13*, 32–44.

4. Petersen, R.H.; Parmasto, E. A redescription of *Gloeostereum incarnatum*. *Mycol. Res.* **1993**, *97*, 1213–1216. [CrossRef]

5. Zhang, Z.-F.; Lv, G.-Y.; Jiang, X.; Cheng, J.-H.; Fan, L.-F. Extraction optimization and biological properties of a polysaccharide isolated from *Gloeostereum incarnatum*. *Carbohydr. Polym.* **2015**, *117*, 185–191. [CrossRef] [PubMed]

6. Wang, D.; Li, Q.; Qu, Y.; Wang, M.; Li, L.; Liu, Y.; Li, Y. The investigation of immunomodulatory activities of *Gloeostereum incaratum* polysaccharides in cyclophosphamide-induced immunosuppression mice. *Exp. Ther. Med.* **2018**, *15*, 3633–3638. [CrossRef] [PubMed]

7. Lull, C.; Wichers, H.J.; Savelkoul, H.F. Antiinflammatory and immunomodulating properties of fungal metabolites. *Mediat. Inflamm.* **2005**, *2005*, 63–80. [CrossRef] [PubMed]

8. Asai, R.; Mitsuhashi, S.; Shigetomi, K.; Miyamoto, T.; Ubukata, M. Absolute configurations of (−)-hirsutanol A and (−)-hirsutanol C produced by *Gloeostereum incarnatum*. *J. Antibiot.* **2011**, *64*, 693–696. [CrossRef] [PubMed]

9. Liu, W.; Chen, L.; Cai, Y.; Zhang, Q.; Bian, Y. Opposite polarity monospore genome *de novo* sequencing and comparative analysis rreveal the possible heterothallic life cycle of *Morchella importuna*. *Int. J. Mol. Sci.* **2018**, *19*, 2525. [CrossRef]

10. Dai, Y.; Su, W.; Yang, C.; Song, B.; Li, Y.; Fu, Y. Development of novel polymorphic EST-SSR markers in Bailinggu (*Pleurotus tuoliensis*) for crossbreeding. *Genes* **2017**, *8*, 325. [CrossRef] [PubMed]

11. Lu, M.Y.; Fan, W.L.; Wang, W.F.; Chen, T.; Tang, Y.C.; Chu, F.H.; Chang, T.T.; Wang, S.Y.; Li, M.Y.; Chen, Y.H.; et al. Genomic and transcriptomic analyses of the medicinal fungus *Antrodia cinnamomea* for its metabolite biosynthesis and sexual development. *Proc. Natl. Acad. Sci. USA* **2014**, *111*, E4743–E4752. [CrossRef] [PubMed]

12. Chen, S.; Xu, J.; Liu, C.; Zhu, Y.; Nelson, D.R.; Zhou, S.; Li, C.; Wang, L.; Guo, X.; Sun, Y.; et al. Genome sequence of the model medicinal mushroom *Ganoderma lucidum*. *Nat. Commun.* **2012**, *3*, 913. [CrossRef] [PubMed]

13. Chen, J.; Zeng, X.; Yang, Y.L.; Xing, Y.M.; Zhang, Q.; Li, J.M.; Ma, K.; Liu, H.W.; Guo, S.X. Genomic and transcriptomic analyses reveal differential regulation of diverse terpenoid and polyketides secondary metabolites in *Hericium erinaceus*. *Sci. Rep.* **2017**, *7*, 10151. [CrossRef] [PubMed]

14. Takazawa, H.; Kashino, S. Incarnal. A new antibacterial sesquiterpene from Basidiomycetes. *Chem. Pharm. Bull.* **1991**, *39*, 555–557. [CrossRef] [PubMed]

15. Li, C.; Lin, F.; An, D.; Wang, W.; Huang, R. Genome sequencing and assembly by long reads in plants. *Genes* **2017**, *9*, 6. [CrossRef] [PubMed]

16. Fu, Y.; Dai, Y.; Yang, C.; Wei, P.; Song, B.; Yang, Y.; Sun, L.; Zhang, Z.W.; Li, Y. Comparative transcriptome analysis identified candidate genes related to Bailinggu mushroom formation and genetic markers for genetic analyses and breeding. *Sci. Rep.* **2017**, *7*, 9266. [CrossRef] [PubMed]

17. Parra, G.; Bradnam, K.; Korf, I. CEGMA: A pipeline to accurately annotate core genes in eukaryotic genomes. *Bioinformatics* **2007**, *23*, 1061–1067. [CrossRef] [PubMed]

18. Simao, F.A.; Waterhouse, R.M.; Ioannidis, P.; Kriventseva, E.V.; Zdobnov, E.M. BUSCO: Assessing genome assembly and annotation completeness with single-copy orthologs. *Bioinformatics* **2015**, *31*, 3210–3212. [CrossRef] [PubMed]

19. Stanke, M.; Keller, O.; Gunduz, I.; Hayes, A.; Waack, S.; Morgenstern, B. AUGUSTUS: Ab initio prediction of alternative transcripts. *Nucleic Acids Res.* **2006**, *34*, W435–W439. [CrossRef] [PubMed]

20. Burge, C.; Karlin, S. Prediction of complete gene structures in human genomic DNA. *J. Mol. Biol.* **1997**, *268*, 78–94. [CrossRef] [PubMed]

21. Allen, J.E.; Majoros, W.H.; Pertea, M.; Salzberg, S.L. JIGSAW, GeneZilla, and GlimmerHMM: Puzzling out the features of human genes in the ENCODE regions. *Genome Biol.* **2006**, *7* (Suppl. 1), S9. [CrossRef]

22. Korf, I. Gene finding in novel genomes. *BMC Bioinform.* **2004**, *5*, 59. [CrossRef] [PubMed]

23. Elsik, C.G.; Mackey, A.J.; Reese, J.T.; Milshina, N.V.; Roos, D.S.; Weinstock, G.M. Creating a honey bee consensus gene set. *Genome Biol.* **2007**, *8*, R13. [CrossRef] [PubMed]

24. Ashburner, M.; Ball, C.A.; Blake, J.A.; Botstein, D.; Butler, H.; Cherry, J.M.; Davis, A.P.; Dolinski, K.; Dwight, S.S.; Eppig, J.T.; et al. Gene ontology: Tool for the unification of biology. The gene ontology consortium. *Nat. Genet.* **2000**, *25*, 25–29. [CrossRef] [PubMed]

25. Tatusov, R.L.; Galperin, M.Y.; Natale, D.A.; Koonin, E.V. The COG database: A tool for genome-scale analysis of protein functions and evolution. *Nucleic Acids Res.* **2000**, *28*, 33–36. [CrossRef] [PubMed]

26. Kanehisa, M.; Goto, S.; Kawashima, S.; Okuno, Y.; Hattori, M. The KEGG resource for deciphering the genome. *Nucleic Acids Res.* **2004**, *32*, D277–D280. [CrossRef] [PubMed]

27. Jurka, J.; Kapitonov, V.V.; Pavlicek, A.; Klonowski, P.; Kohany, O.; Walichiewicz, J. Repbase Update, a database of eukaryotic repetitive elements. *Cytogenet. Genome Res.* **2005**, *110*, 462–467. [CrossRef] [PubMed]

28. Tarailo-Graovac, M.; Chen, N. Using RepeatMasker to identify repetitive elements in genomic sequences. *Curr. Protoc. Bioinform.* **2009**. [CrossRef]

29. Benson, G. Tandem repeats finder: A program to analyze DNA sequences. *Nucleic Acids Res.* **1999**, *27*, 573–580. [CrossRef] [PubMed]

30. Lagesen, K.; Hallin, P.; Rodland, E.A.; Staerfeldt, H.H.; Rognes, T.; Ussery, D.W. RNAmmer: Consistent and rapid annotation of ribosomal RNA genes. *Nucleic Acids Res.* **2007**, *35*, 3100–3108. [CrossRef] [PubMed]

31. Lowe, T.M.; Eddy, S.R. tRNAscan-SE: A program for improved detection of transfer RNA genes in genomic sequence. *Nucleic Acids Res.* **1997**, *25*, 955–964. [CrossRef] [PubMed]

32. Gardner, P.P.; Daub, J.; Tate, J.G.; Nawrocki, E.P.; Kolbe, D.L.; Lindgreen, S.; Wilkinson, A.C.; Finn, R.D.; Griffiths-Jones, S.; Eddy, S.R.; et al. Rfam: Updates to the RNA families database. *Nucleic Acids Res.* **2009**, *37*, D136–D140. [CrossRef] [PubMed]

33. Edgar, R.C. MUSCLE: Multiple sequence alignment with high accuracy and high throughput. *Nucleic Acids Res.* **2004**, *32*, 1792–1797. [CrossRef] [PubMed]

34. Stamatakis, A. RAxML version 8: A tool for phylogenetic analysis and post-analysis of large phylogenies. *Bioinformatics* **2014**, *30*, 1312–1313. [CrossRef] [PubMed]

35. Yang, Z. PAML 4: Phylogenetic analysis by maximum likelihood. *Mol. Biol. Evol.* **2007**, *24*, 1586–1591. [CrossRef] [PubMed]

36. Cantarel, B.L.; Coutinho, P.M.; Rancurel, C.; Bernard, T.; Lombard, V.; Henrissat, B. The Carbohydrate-Active EnZymes database (CAZy): An expert resource for Glycogenomics. *Nucleic Acids Res.* **2009**, *37*, D233–D238. [CrossRef] [PubMed]

37. Nelson, D.R. The cytochrome p450 homepage. *Hum. Genom.* **2009**, *4*, 59–65.

38. Weber, T.; Blin, K.; Duddela, S.; Krug, D.; Kim, H.U.; Bruccoleri, R.; Lee, S.Y.; Fischbach, M.A.; Muller, R.; Wohlleben, W.; et al. antiSMASH 3.0—A comprehensive resource for the genome mining of biosynthetic gene clusters. *Nucleic Acids Res.* **2015**, *43*, W237–W243. [CrossRef] [PubMed]

39. Trapnell, C.; Pachter, L.; Salzberg, S.L. TopHat: Discovering splice junctions with RNA-Seq. *Bioinformatics* **2009**, *25*, 1105–1111. [CrossRef] [PubMed]

40. Anders, S.; Pyl, P.T.; Huber, W. HTSeq—A Python framework to work with high-throughput sequencing data. *Bioinformatics* **2015**, *31*, 166–169. [CrossRef] [PubMed]

41. Anders, S.; Huber, W. Differential expression analysis for sequence count data. *Genome Biol.* **2010**, *11*, R106. [CrossRef] [PubMed]

42. Shim, D.; Park, S.G.; Kim, K.; Bae, W.; Lee, G.W.; Ha, B.S.; Ro, H.S.; Kim, M.; Ryoo, R.; Rhee, S.K.; et al. Whole genome de novo sequencing and genome annotation of the world popular cultivated edible mushroom, *Lentinula edodes*. *J. Biotechnol.* **2016**, *223*, 24–25. [CrossRef] [PubMed]

43. Gupta, D.K.; Ruhl, M.; Mishra, B.; Kleofas, V.; Hofrichter, M.; Herzog, R.; Pecyna, M.J.; Sharma, R.; Kellner, H.; Hennicke, F.; et al. The genome sequence of the commercially cultivated mushroom *Agrocybe aegerita* reveals a conserved repertoire of fruiting-related genes and a versatile suite of biopolymer-degrading enzymes. *BMC Genom.* **2018**, *19*, 48. [CrossRef] [PubMed]

44. Shu, S.; Chen, B.; Zhou, M.; Zhao, X.; Xia, H.; Wang, M. De novo sequencing and transcriptome analysis of *Wolfiporia cocos* to reveal genes related to biosynthesis of triterpenoids. *PLoS ONE* **2013**, *8*, e71350. [CrossRef] [PubMed]

45. Morin, E.; Kohler, A.; Baker, A.R.; Foulongne-Oriol, M.; Lombard, V.; Nagye, L.G.; Ohm, R.A.; Patyshakuliyeva, A.; Brun, A.; Aerts, A.L.; et al. Genome sequence of the button mushroom *Agaricus bisporus* reveals mechanisms governing adaptation to a humic-rich ecological niche. *Proc. Natl. Acad. Sci. USA* **2012**, *109*, 17501–17506. [CrossRef] [PubMed]

46. Yap, H.Y.; Chooi, Y.H.; Firdaus-Raih, M.; Fung, S.Y.; Ng, S.T.; Tan, C.S.; Tan, N.H. The genome of the Tiger Milk mushroom, *Lignosus rhinocerotis*, provides insights into the genetic basis of its medicinal properties. *BMC Genom.* **2014**, *15*, 635. [CrossRef] [PubMed]
47. Xiao, D.; Ma, L.; Yang, C.; Ying, Z.; Jiang, X.; Lin, Y.Q. De novo sequencing of a *Sparassis latifolia* genome and its associated comparative analyses. *Can. J. Infect. Dis. Med. Microbiol.* **2018**, *2018*, 1857170. [CrossRef] [PubMed]
48. Kurata, A.; Fukuta, Y.; Mori, M.; Kishimoto, N.; Shirasaka, N. Draft genome sequence of the basidiomycetous fungus *Flammulina velutipes* TR19. *Genome Announc.* **2016**, *4*. [CrossRef] [PubMed]
49. Zhong, J.J.; Xiao, J.H. Secondary metabolites from higher fungi: Discovery, bioactivity, and bioproduction. *Adv. Biochem. Eng. Biotechnol.* **2009**, *113*, 79–150. [CrossRef] [PubMed]
50. Floudas, D.; Binder, M.; Riley, R.; Barry, K.; Blanchette, R.A.; Henrissat, B.; Martinez, A.T.; Otillar, R.; Spatafora, J.W.; Yadav, J.S.; et al. The Paleozoic origin of enzymatic lignin decomposition reconstructed from 31 fungal genomes. *Science* **2012**, *336*, 1715–1719. [CrossRef] [PubMed]
51. Sista Kameshwar, A.K.; Qin, W. Comparative study of genome-wide plant biomass-degrading CAZymes in white rot, brown rot and soft rot fungi. *Mycology* **2018**, *9*, 93–105. [CrossRef] [PubMed]
52. Martinez, A.T.; Ruiz-Duenas, F.J.; Martinez, M.J.; Del Rio, J.C.; Gutierrez, A. Enzymatic delignification of plant cell wall: From nature to mill. *Curr. Opin. Biotechnol.* **2009**, *20*, 348–357. [CrossRef] [PubMed]
53. Lee, H.Y.; Moon, S.; Shim, D.; Hong, C.P.; Lee, Y.; Koo, C.D.; Chung, J.W.; Ryu, H. Development of 44 novel polymorphic SSR markers for determination of shiitake mushroom (*Lentinula edodes*) cultivars. *Genes* **2017**, *8*, 109. [CrossRef] [PubMed]
54. Gao, J.; Yue, D.C.; Cheng, K.D.; Wang, S.C.; Yu, K.B.; Zheng, Q.T.; Yang, J.S. Gloeosteretriol, a new sesquiterpene from the fermentation products of *Gloeostereum incarnatum* S. Ito et Imai. *Yao Xue Xue Bao = Acta Pharm. Sin.* **1992**, *27*, 33–36.
55. Li, H.J.; Chen, T.; Xie, Y.L.; Chen, W.D.; Zhu, X.F.; Lan, W.J. Isolation and structural elucidation of chondrosterins F-H from the marine fungus *Chondrostereum* sp. *Mar. Drugs* **2013**, *11*, 551–558. [CrossRef] [PubMed]
56. Christianson, D.W. Unearthing the roots of the terpenome. *Curr. Opin. Chem. Biol.* **2008**, *12*, 141–150. [CrossRef] [PubMed]
57. Hidalgo, P.I.; Ullan, R.V.; Albillos, S.M.; Montero, O.; Fernandez-Bodega, M.A.; Garcia-Estrada, C.; Fernandez-Aguado, M.; Martin, J.F. Molecular characterization of the PR-toxin gene cluster in *Penicillium roqueforti* and *Penicillium chrysogenum*: Cross talk of secondary metabolite pathways. *Fungal Genet. Biol. FG B* **2014**, *62*, 11–24. [CrossRef] [PubMed]
58. Schothorst, R.C.; van Egmond, H.P. Report from SCOOP task 3.2.10 "collection of occurrence data of Fusarium toxins in food and assessment of dietary intake by the population of EU member states". Subtask: Trichothecenes. *Toxicol. Lett.* **2004**, *153*, 133–143. [CrossRef] [PubMed]
59. Benveniste, P. Biosynthesis and accumulation of sterols. *Annu. Rev. Plant Biol.* **2004**, *55*, 429–457. [CrossRef] [PubMed]
60. Rios, J.L.; Andujar, I.; Recio, M.C.; Giner, R.M. Lanostanoids from fungi: A group of potential anticancer compounds. *J. Nat. Prod.* **2012**, *75*, 2016–2044. [CrossRef] [PubMed]
61. Schmidt-Dannert, C. Biosynthesis of terpenoid natural products in fungi. *Adv. Biochem. Eng. Biotechnol.* **2015**, *148*, 19–61. [CrossRef] [PubMed]
62. Kanehisa, M.; Goto, S. KEGG: Kyoto encyclopedia of genes and genomes. *Nucleic Acids Res.* **2000**, *28*, 27–30. [CrossRef] [PubMed]
63. Cresnar, B.; Petric, S. Cytochrome P450 enzymes in the fungal kingdom. *Biochim. Biophys. Acta* **2011**, *1814*, 29–35. [CrossRef] [PubMed]
64. Sanglard, D.; Loper, J.C. Characterization of the alkane-inducible cytochrome P450 (P450alk) gene from the yeast *Candida tropicalis*: Identification of a new P450 gene family. *Gene* **1989**, *76*, 121–136. [CrossRef]
65. Mansuy, D. The great diversity of reactions catalyzed by cytochromes P450. *Comp. Biochem. Physiol. Part C Pharmacol. Toxicol. Endocrinol.* **1998**, *121*, 5–14. [CrossRef]
66. Park, J.; Lee, S.; Choi, J.; Ahn, K.; Park, B.; Park, J.; Kang, S.; Lee, Y.H. Fungal cytochrome P450 database. *BMC Genom.* **2008**, *9*, 402. [CrossRef] [PubMed]
67. Xu, Z.; Chen, X.; Zhong, Z.; Chen, L.; Wang, Y. *Ganoderma lucidum* polysaccharides: Immunomodulation and potential anti-tumor activities. *Am. J. Chin. Med.* **2011**, *39*, 15–27. [CrossRef] [PubMed]

68. Montijn, R.C.; Vink, E.; Muller, W.H.; Verkleij, A.J.; Van Den Ende, H.; Henrissat, B.; Klis, F.M. Localization of synthesis of beta1,6-glucan in *Saccharomyces cerevisiae*. *J. Bacteriol.* **1999**, *181*, 7414–7420. [PubMed]
69. Yuan, Y.; Wu, F.; Si, J.; Zhao, Y.F.; Dai, Y.C. Whole genome sequence of *Auricularia heimuer* (Basidiomycota, Fungi), the third most important cultivated mushroom worldwide. *Genomics* **2017**. [CrossRef] [PubMed]

Article

Genome Sequencing of *Cladobotryum protrusum* Provides Insights into the Evolution and Pathogenic Mechanisms of the Cobweb Disease Pathogen on Cultivated Mushroom

Frederick Leo Sossah [1,†], Zhenghui Liu [2,†], Chentao Yang [3,†], Benjamin Azu Okorley [1], Lei Sun [1], Yongping Fu [1,*] and Yu Li [1,*]

[1] Engineering Research Center of Chinese Ministry of Education for Edible and Medicinal Fungi, Jilin Agricultural University, Changchun 130118, China; flsossah@gmail.com (F.L.S.); bazu_okorley@st.ug.edu.gh (B.A.O.); sunlei@jlau.edu.cn (L.S.)

[2] Department of Plant Protection, Shenyang Agricultural University, Shenyang 110866, China; zhenghuiliu1212@126.com

[3] BGI-Shenzhen, Shenzhen 518083, China; China National GeneBank, BGI-Shenzhen, Shenzhen 518083, China; yangchentao@genomics.cn

* Correspondence: yongpingfu81@126.com (Y.F.); yuli966@126.com (Y.L.); Tel.: +86-431-8453-2989.

† These authors contributed equally to this work.

Received: 15 January 2019; Accepted: 5 February 2019; Published: 8 February 2019

Abstract: *Cladobotryum protrusum* is one of the mycoparasites that cause cobweb disease on cultivated edible mushrooms. However, the molecular mechanisms of evolution and pathogenesis of *C. protrusum* on mushrooms are largely unknown. Here, we report a high-quality genome sequence of *C. protrusum* using the single-molecule, real-time sequencing platform of PacBio and perform a comparative analysis with closely related fungi in the family Hypocreaceae. The *C. protrusum* genome, the first complete genome to be sequenced in the genus *Cladobotryum*, is 39.09 Mb long, with an N50 of 4.97 Mb, encoding 11,003 proteins. The phylogenomic analysis confirmed its inclusion in Hypocreaceae, with its evolutionary divergence time estimated to be ~170.1 million years ago. The genome encodes a large and diverse set of genes involved in secreted peptidases, carbohydrate-active enzymes, cytochrome P450 enzymes, pathogen–host interactions, mycotoxins, and pigments. Moreover, *C. protrusum* harbors arrays of genes with the potential to produce bioactive secondary metabolites and stress response-related proteins that are significant for adaptation to hostile environments. Knowledge of the genome will foster a better understanding of the biology of *C. protrusum* and mycoparasitism in general, as well as help with the development of effective disease control strategies to minimize economic losses from cobweb disease in cultivated edible mushrooms.

Keywords: *Cladobotryum protrusum*; mycoparasite; cobweb disease; *de novo* assembly; SMRT sequencing

---

1. Introduction

As the commercial cultivation of edible mushrooms continuously expands worldwide, the occurrence of diseases caused by fungal pathogens is also increasing, seriously affecting mushroom quality and yield [1]. Cobweb disease is one of the most important limiting factors in mushroom production [2]. Members of the genus *Cladobotryum*, belonging to the kingdom Fungi, division Ascomycota, class Sordariomycetes, order Hypocreales, and family Hypocreaceae, are causal agents of cobweb disease on a number of economically important mushroom crops, such as *Agaricus bisporus*, *Flammulina velutipes*, *Pleurotus ostreatus*, *P. eryngii*, *Hypsizygus marmoreus*, and *Ganoderma lucidum* [3–8]. The species *C. dendroides*, *C. mycophilum*, *C. protrusum*, and *C. varium* are pathogens that frequently cause cobweb disease in commercial mushroom farms. The characteristic symptom of cobweb disease is the abundance of coarse mycelium [9], which covers the affected mushrooms with numerous spores and spreads rapidly in commercial growth rooms, leading to serious economic losses worldwide [5,10,11].

Among the *Cladobotryum* genus, *C. protrusum* is an important member, as it causes cobweb disease on edible mushrooms, such as *Coprinus comatus*, *Agaricus bisporus*, and *P. ostreatus*, and has the widest distribution [12,13]. The taxonomy, classification, incidence, distribution, and host range of *C. protrusum* have been well studied [12,13]. The phylogenetic placement of *C. protrusum* within the genus *Cladobotryum* has been inferred from the internal transcribed spacer (ITS), translation-elongation factor 1-alpha, and DNA-directed RNA polymerase II subunit *RPB1* and *RPB2* genes [13]. Beyond this study, no genetic resources of *C. protrusum* have been developed. Specifically, the infection mechanism of mycoparasitism is largely unknown, and, in particular, the genes related to pathogenicity, virulence, cell wall degrading enzymes, and secondary metabolites (SMs) are undetermined. Therefore, the sequenced genome could serve as an important genetic resource for further evolutionary studies of the *Cladobotryum* genus and facilitate the elucidation of the pathogenic mechanisms of *C. protrusum*.

The *Cladobotryum* genus comprises at least 66 species [12], and genome sequencing has not been performed on any of them. The development of next-generation sequencing technologies, such as Illumina, 454 sequencing platforms, and the single-molecule real-time (SMRT, PacBio) sequencing platform, has led to the sequencing of many fungal genomes [13]. PacBio sequencing technology offers increased read lengths, unbiased genome coverage, and simultaneous identification of mutation sites [14–16]. Sequenced genomes provide data that allow us to gain insights into fungal growth, evolution, and host–pathogen interactions as well as identifying genes related to pathogenicity and the synthesis of SMs of economic importance [17].

In this study, we report the de novo genome sequencing of *C. protrusum* generated using the SMRT sequencing platform, which is the first genome to be sequenced in the *Cladobotryum* genus. We aim (1) to present a high-quality reference genome for *C. protrusum* and an analysis of genes related to its pathogenicity and mycoparasitism and (2) to conduct a comparative genome analysis using other sequenced genomes from species within the Hypocreaceae family. The genome assembly will further expand genomic datasets for comparative genomics of species in the Hypocreaceae family and mycoparasites in general. This study will promote the understanding of the biology of *C. protrusum* and the development of effective strategies for controlling cobweb disease.

2. Materials and Methods

2.1. Fungal Strain and Genomic DNA Extraction

The *C. protrusum* strain used in this study was a single spore isolate collected from the Institute of Applied Mycology, Huazhong Agricultural University, Wuhan, Hubei, China, which was maintained on potato dextrose agar (Difco™, Fisher Scientific, Pittsburgh, PA, USA). The fungal strain was isolated from *C. comatus* from a mushroom farm in Wuhan [18]. The identity of the fungus was confirmed by morphological characteristics, PCR amplification, and sequencing of the ITS gene sequence of the genomic DNA and a BLAST search on the GenBank database. Mycelium plugs of pure isolates were cultured on PDA overlaid with cellophane membrane and incubated at 25 °C for three days under a 12 h white light photoperiod. Genomic DNA was extracted from mycelia using the CWBiotech Plant DNA extraction kit (CWBiotech Corporation, Beijing China) following the manufacturer's instructions. The quality of DNA was verified with 1% agarose gel electrophoresis and visualization with Gel Doc™ XR+ (Bio-Rad, USA). DNA quantification was done using a Qubit 4.0 fluorometer (Invitrogen, Carlsbad, CA, USA).

2.2. Genome Sequencing and Assembly

A genomic DNA library was constructed using a SMRTbell Template Prep kit (Pacific Biosciences, CA, USA) in accordance with the manufacturer's protocol. A BluePippin device (Sage Science,

Inc., Beverly, MA, USA) was used to select 20 kb insert size fragments for the SMRTbell Template library. Quality inspection and quantification of the size-selected library were done using an Agilent 2100 Bioanalyzer (Agilent Technologies, Santa Clara, CA, USA) and Qubit 4.0 fluorometer (Invitrogen, Carlsbad, CA, USA). Prepared whole-genome libraries were sequenced on a PacBio Sequel sequencer (Pacific Biosciences, Menlo Park, CA, USA) with one SMRT cell at the Engineering Research Center of the Chinese Ministry of Education for Edible and Medicinal Fungi, Jilin Agricultural University, Changchun, China. The genome was assembled using SMARTdenovo as described below, in accordance with www.github.com/smartdenovo. The completeness of the assembled genome was evaluated using the Core Eukaryotic Genes Mapping Approach (CEGMA) [19] and Benchmarking Universal Single-Copy Orthologs (BUSCO v3) [20,21] with conserved orthologous gene profiles for fungi.

2.3. Gene Prediction and Annotations

The assembled genome was annotated using a homology-based method and de novo prediction methods. Genewise [22] was used for the homology search using the proteomes of *Fusarium redolens*, *Fusarium oxysporum* FOX64, *Neurospora crassa*, and *Trichoderma atroviride* (available from http://www.uniprot.org/; release 2012_07) as training sets. De novo prediction of the protein-coding genes was done using Augustus v2.7 [23], GlimmerHMM v3.02 [24], Genscan v1.0 [25], and SNAP v 2006-07-28 [26]. GLEAN was used to integrate all of the gene models to produce a non-redundant reference gene set (http://glean-gene.sourceforge.net/) [27]. The repeat sequences were identified and masked using RepeatModeler v1.0.7 and RepeatMasker v4.0.5. Tandem repeats were identified by the Tandem Repeats Finder (TRF) v4.04 (http://tandem.bu.edu/trf/trf.html) [28] by searching the repeat sequences against the Repbase database (http://www.girinst.org/repbase/) [29]. Transfer RNAs were predicted using tRNAscan-SE 1.3.1 [30], whereas rRNAs and noncoding RNAs were identified using RNAmmer 1.2 [31] and the Rfam database [32]. The predicted-coding sequences were functionally annotated by BLASTP (e-value cutoff of 1×10^{-5}) query against several protein databases such as the National Center for Biotechnology Information (NCBI) non-redundant (nr), Cluster of Orthologous Groups (COG) [33], the Gene Ontology (GO) database [34], the Kyoto Encyclopedia of Genes and Genomes (KEGG) database [35], the SwissProt database [36–38], the TrEMBL databases [38], and the InterPro Protein Families Database (IPR including Pfam database) [39]. The mating-type genes for *C. protrusum* were determined by BLAST (tBLASTx e-value 1×10^{-30}) similarity searches using mating-type genes and flanking gene sequences from the order Hypocreales retrieved from NCBI database. The gene structure was drawn using the software package illustrator of biological sequences version 1.0 (http://ibs.biocuckoo.org/) [40].

2.4. Orthologous Gene Families and Phylogenomic Analysis

An all-vs.-all BLASTP with an e-value cutoff of 1×10^{-5} was used to compare the proteins of ten species including *C. protrusum* (CPR), *Clonostachys rosea* (CR), *Fusarium solani* (FS), *Magnaporthe grisea* (MG), *Metarhizium acridum* (MA), *N. crassa* (NA), *Tolypocladium inflatum* (TI), *T. longibrachiatum* (TL), *T. reesei* (TR), and *T. virens* (TV) (Supplementary Table S1). The BLAST results were clustered by a MATLAB implementation of the Markov Clustering (MCL) algorithm (MMCL) [41] to identify orthologous groups using OrthoMCL (v. 2.0.9) [42]. Multiple sequence alignment was performed on the proteins of single-copy orthologs identified using MUSCLE [43]. The phylogenetic tree was used for maximum-likelihood (ML) analysis by RAxML-8.0.26 [44] using the LG+I+G+F amino acid substitution matrix model selected by ProtTest (v. 3.4) [45] with 1000 bootstrap replicates.

The species divergence times were inferred with the MCMCTree included in the PAML v4.7a software package [46] with r8s v1.81 (http://loco.biosci.arizona.edu/r8s/) [47]. The divergence times were estimated using the approximate method with fossil calibrations from http://www.timetree.org [48]. The expansion of the orthologous gene families and contraction across organisms was calculated by Computational Analysis of Gene Family Evolution (CAFE) (v. 3) [49] with a stochastic birth and death model using a lambda value of 0.314, a *p*-value of 0.01, and 1000 random samples [50]. The genes under selection pressure were identified by calculating the dN/dS ratio between the species in the phylogenetic tree ($p \leq 0.01$) using the Codeml program PAML [46].

Furthermore, OrthoVenn [51] was used for genome-wide identification, comparison, and visualization of unique and shared orthologous gene clusters for *C. protrusum*, *Escovopsis weberi*, *T. reesei*, and *T. virens*. In addition, the proteomes of *C. protrusum*, *E. weberi*, *T. reesei*, *T. virens*, *M. grisea*, and *Aspergillus nidulans* were clustered into orthologous groups using OrthoFinder [52]. The multiple sequence alignments of the single-copy orthologs was used for phylogenetic analysis using the Neighbor-Joining method, which was conducted in MEGAX, to validate the relationships among *C. protrusum* and the other three fungi in the family Hypocreaceae [53].

2.5. Secretory Protein Analysis and Pathogenicity-Related Genes

Secretory proteins were predicted using SignalP 3.0 [54]. Transmembrane proteins were predicted by TMHMM [55]. Protein localization signals, excluding those related to the plastid location, were identified using TargetP [56]. Glycosylphosphatidylinositol (GPI)-anchored proteins were predicted using the PredGPI server [57]. Transporters were analyzed through local BLASTP against the Transporter Classification Database (TCdb) with a cutoff e-value of 1×10^{-40} [58]. Proteases were identified with BLASTP (e-value 1×10^{-30}) by searching the secretory proteins against the MEROPS database [59]. Cytochrome P450s were classified based on BLASTP alignment against the P450 database with a cutoff e-value of 1×10^{-20} (http://drnelson.uthsc.edu/CytochromeP450.html) [60]. To identify virulence-associated genes, BLASTP (with a cutoff e-value of 1×10^{-5}) searches of the *C. protrusum* genome were performed against protein sequences in the pathogen–host interaction database (PHI) (version 3.2, http://www.phi-base.org/) [61] and the database of fungal virulence factors (DFVF) [62]. Carbohydrate-active enzymes (CAZymes) were determined using the dbCAN 2 meta server [63]. SMs were annotated using the antiSMASH (http://antismash.secondarymetabolites.org) fungiSMASH option [64] database and NaPDoS (http://napdos.ucsd.edu) [65].

3. Results

3.1. Genome Sequencing and Assembly of C. protrusum

The genome of *C. protrusum* was sequenced using the PacBio SMRT Sequel platform. In total, 587,476 sub-reads were generated, representing 6.23 Gb of sequence data at 160 X coverage. The de novo assembly of the *C. protrusum* genome yielded ~39.09 Mb, consisting of 18 scaffolds (Table 1) with a scaffold N50 length of 4.97 Mbp and a scaffold N90 length of 1.93 Mbp. The guanine-cytosine content (GC-content) of the *C. protrusum* genome was 47.84%. CEGMA [19] analysis revealed that 97.58% of the core eukaryotic genes were contained in the assembly (242 out of 248 core eukaryotic genes), while the BUSCO [21] assessment results showed that 99.7% (289 out of 290 genes) of genes were covered by the assembled genome containing 99%, 0.7%, and 0.3% of complete, duplicated, and missing BUSCOs [21], respectively. Therefore, the CEGMA [19] and BUSCO [21] results indicate that the assembled genome for *C. protrusum* was of a high quality. The genome of *C. protrusum* has been deposited into the NCBI database with the accession number RZGP00000000.

Table 1. The genome features of *C. protrusum*.

Genome Features	*C. protrusum*
Genome size (Mb)	39.087
Total number of scaffolds	18
Total length of scaffold sequences (Assembly size)	39,087,229 bp
Scaffold N50	4,973,539 bp
Scaffold N90:	1,928,814 bp
GC-content (%):	47.84%
N Length:	0bp
N content (%):	0.0%
Transposable elements (%)	2.59
Predicted proteins	11,003
tRNA	242
rRNA	225
miRNA	97
snRNA	22

3.2. Genome Annotation

Genome annotation based on de novo prediction and a homology-based search identified 11,003 protein-coding genes with an average sequence length of 1723.49 bp (Table 1). Overall, 10,623 (96.55%) of the predicted genes had known homologs in at least one functional protein database. Among these proteins, 10,607 (96.40%) were similar to the sequences in the NCBI nr database, 6899 (62.70%) homologs were similar to sequences in Swiss-Prot, 6786 (61.67%) were mapped to KEGG, 4895 (44.49%) were classified in COG, 10,587 (96.22%) were classified in TrEMBL, 7184 (65.29%) were classified in InterPro, and 5332 (48.46%) were assigned to GO terms (Figure 1A). In addition, the proportion of transposable elements (TEs) in *C. protrusum* was estimated to be 2.59% based on combined homology-based and de novo approaches (Table 1). The TEs were randomly distributed across all chromosomes, and the Class I TEs (retrotransposons) (1.34%) were more abundant than the Class II TEs (DNA transposons) (0.48%). The unknown TEs represented 1.24% of the total, and the most abundant characterized TEs in the *C. protrusum* genome were long terminal repeat (LTR) retrotransposons, which accounted for 0.67% of the genome (Supplementary Table S2). A total of 242 tRNAs and 225 rRNAs of the non-coding RNA species were identified in the *C. protrusum* genome. We also predicted 97 miRNAs and 22 snRNAs.

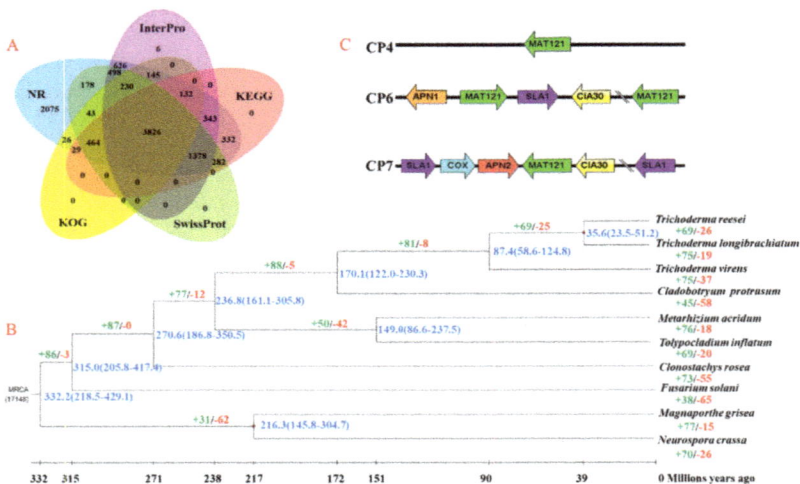

Figure 1. Annotation, phylogenetic and divergence time tree, and mating-type gene structure of the *C. protrusum* genome assembly. (**A**) Functional annotation of the protein-coding genes in the

C. protrusum genome. (**B**) Phylogenetic and divergence time tree of *C. protrusum* and other nine fungal species. The phylogenetic tree was generated from 3279 single-copy orthologs using the maximum-likelihood method. The divergence time range is shown in blue text, the numbers in green/ red show the proportion of expanded/contracted gene families in each fungal species. (**C**) Schematic representation of the structure of mating-type loci (*MAT 1-2-1*) in *C. protrusum*. The arrows represent the orientation of the *MAT1-2* genes *SLA*, *APN*, *CIA30*, and *COX*.

3.3. Identification of Mating-Type Idiomorphs in C. protrusum

MAT1-2 mating-type idiomorphs were identified in the genome of *C. protrusum*, whereas the *MAT1-1* idiomorph (1α domains) was not. The *MAT1-2* idiomorphs were located on different scaffolds (utg37, utg67 (two genes), and utg83) and were distant from each other (Figure 1C). This result suggests that *C. protrusum* has tetrapolar nuclei and confirms our previous observations under a microscope, which showed four nuclei. The cytoskeleton assembly control protein, AP endonuclease, cytochrome C oxidase subunit VIa, and complex I intermediate-associated protein 30 kDa genes were found to flank the *MAT1-2* idiomorph on utg67 and utg83.

3.4. Genome Evolution and Phylogenomic Analysis of C. protrusum

A total of 122,201 genes from ten species, including CPR, CR, FS, MG, MA, NA, TI, TL, TR, and TV, were clustered into 11,976 orthogroups using OrthoMCL. Among them, 4761 (39.75%) gene families were shared among all ten species, while 3279 (27.38%) were single-copy orthologous genes. A total of 862 (7.20%) gene families were found to be unique to *C. protrusum* when compared to the other genomes. The single-copy orthologous genes were used for the phylogenetic analysis of the above-mentioned ten species, which was conducted to determine the relationship between *C. protrusum* and other important members in the class Sordariomycetes (Figure 1B). The phylogenetic analysis resolved the ten species into three orders—Hypocreales, Magnaporthales, and Sordariales—with five families in Hypocreales clustered in a different node with NA and MG under separate nodes as outgroups. The orders Hypocreales, Magnaporthales, and Sordariales diverged from the most recent common ancestor (MRCA) 332.2 million years ago (MYA). *C. protrusum* clustered with *Trichoderma* spp. and was separated into different clades based on the genus. The phylogenetic tree confirmed that *C. protrusum* belongs to the Hypocreaceae family and diverged from the genus *Trichoderma* about 170.1 MYA. The results indicate that *C. protrusum* and *Trichoderma* spp. are distantly related to each other at the family level.

The expansion and contraction of the analysis of gene families showed that 88 (2.68%) gene families expanded and five gene families contracted in the family Hypocreaceae based on the 3279 shared gene families from the phylogenetic tree (Figure 1B). Furthermore, we found that *C. protrusum* gained 45 gene families and lost 58 (1.77%) gene families. Except for *C. protrusum* and *F. solani,* the gain of gene families occurred more often than gene loss in the species analyzed. The expanded gene families contain 245 (7.47%) genes (Supplementary Table S3) with several genes related to metabolism, transcription, proteins with binding functions, signal transduction mechanisms, cell rescue and defense protein transport, and synthesis of SMs. Moreover, the gene families exhibiting the largest expansions in *C. protrusum* include zinc-binding dehydrogenase transcription factors, major facilitator superfamily (MFS) transporters, alcohol dehydrogenases, ankyrin-repeat proteins, ATP-binding cassette (ABC) transporters, polyketide synthases (PKSs), and cytochrome P450 monooxygenases (CYP). Interestingly, genes involved in the mediation and regulation of SM synthesis were the most abundant and included genes such as acyl transferase domain, AMP-binding enzyme, beta-ketoacyl synthase, C-terminal domain, condensation domain, insecticide toxin TCdb middle, methyltransferase domain, polyketide synthase dehydratase, and keto-reductase domain. We also found the vegetative incompatibility or heterokaryon incompatibility protein (HET) in the *C. protrusum* genome.

We identified a total of 3196 genes under selection pressure among the species. Out of these 3196 genes, 14.05% (449 genes) and 24.50% (783 genes) were under positive selection in *C. protrusum* at *p*-values of $p < 0.01$ and $p < 0.05$ respectively (false discovery rate, FDR < 0.1). The positive selection genes (PSGs) were functionally annotated in the GO, KEGG, Pfam, and SwissProt databases (Supplementary Table S4, and Figures S1 and S2). The most abundant GO terms for PSGs were related to cellular component (Figure S1), and, of these, cell (154), cell part (152), and organelle (120) were the three most common GO terms. The PSGs were subsequently analyzed for enrichment in GO categories and KEGG pathways. The analysis revealed 64 enriched metabolic KEGG pathways (Supplementary Table S5), whereas no GO terms were enriched for the PSGs. Further analysis of the PSGs showed that 52 (11.58%) genes are involved in PHI and the majority of the PSGs that played roles in mycoparasitism, include CYP, MFS, SMs, the glycosyl hydrolases family, peptidases, lipases, the subtilase family, and transcription factors.

3.5. The Orthologous Genes of C. protrusum and Three Other Fungi in the Hypocreaceae Family

A phylogenetic tree was constructed using the single-copy orthologs from the clustered proteomes of *C. protrusum*, *E. weberi*, *T. reesei*, *T. virens*, *M. grisea*, and *A. nidulans*. The tree (Figure S3) depicts the relationships among *C. protrusum* and the other three fungi in the family Hypocreaceae. *M. grisea* and *A. nidulans* were used as outgroups in the tree. We used OrthoVenn [51] to cluster orthologous genes and compared the proteomes of *C. protrusum* against *E. weberi*, *T. reesei*, and *T. virens*, (Supplementary Table S1) which belong to the same family, i.e., Hypocreaceae. The species formed 9682 orthologous clusters and 9357 (96.64%) clusters for at least two species. Among them, 5756 (59.45%) orthologous clusters were shared among all four species (Figure 2A). The top three Swiss-Prot annotations among the core shared orthologous proteins include the ATP-binding cassette transporter (13 proteins), the F-box protein (11 proteins), and the Leptomycin B resistance protein (9 proteins) (Supplementary Table S6). The unique orthologous clusters are 168 (1.74%), 5 (0.05%), 9 (0.09%), and 148 (1.53%) for *C. protrusum*, *E. weberi*, *T. reesei*, and *T. virens*, respectively. Similarly, *C. protrusum* had the most enriched GO categories (23) followed by *T. virens* (9), while *E. weberi* and *T. reesei* had no known annotations or GO enrichment (Supplementary Table S7). Most of the unique genes to *C. protrusum* are related to SM biosynthesis. There were 521 (5.38%) gene clusters shared between *C. protrusum* and *T. virens*, 185 (1.91%) for *C. protrusum* and *E. weberi*, and 55 (0.57%) for *C. protrusum* and *T. reesei*. The highest gene cluster shared between any two species with the most enriched GO categories was observed for *T. reesei* and *T. virens*. This could be because they belong to the same genus. The gene clusters of the enriched GO for *T. reesei* and *T. virens* as well as *C. protrusum* and *T. reesei* were related to transport and cell enzyme degradation. The gene clusters of the enriched GO for *C. protrusum* and *E. weberi* were SMs, especially genes related to toxins and pigmentation, e.g., emodin and asperthecin.

A / **B** (Venn diagram and stacked bar chart)

C — Abundance of CAZyme modules

CP	EW	TR	TV	Family
18	5	15	20	AA3
45	12	19	36	AA7
1	0	0	0	AA13
4	3	3	4	CE4
6	2	4	6	CE5
35	14	22	41	CE10
16	13	16	16	GH16
29	16	19	31	GH18
3	2	3	3	GH20
14	11	13	17	GH3
9	8	8	8	GH47
8	5	7	11	GH5
7	4	6	10	GH55
0	0	0	1	GH6
0	1	2	2	GH7
4	1	4	5	GH71
5	1	3	5	GH75
8	6	8	9	GH76
8	5	7	7	GH92
7	8	8	10	GT2
1	1	1	1	GT33
1	0	0	0	GT54
0	2	0	0	PL1
2	0	2	2	PL20
4	1	2	3	PL7
1	1	1	1	PL8

D — NaPDoS SMs

CP	EW	TR	TV	
2	2	2	2	Fatty acid synthesis
0	0	2	1	Microcystin
2	1	0	1	Mycosubtilin
1	0	0	0	Salinisporamide
0	0	1	1	Actinomycin
2	0	2	2	Bacitracin
1	0	6	7	Calcium-dependent antibiotic
4	0	0	0	Complestatin
2	1	1	1	Cyclomarin
16	1	0	13	Cyclosporin
5	0	0	1	Lychenicin
1	0	0	0	Surfactin
2	0	0	1	Syringomycin
0	1	0	0	Thiocoraline
1	0	0	1	Tyrocidin
1	0	0	0	Avermectin
1	0	0	0	Bikaverin
1	0	0	0	Calicheamicin
10	0	2	5	Compactin
2	0	0	1	Epothilone
24	3	3	7	Fumonisin
0	0	1	0	Leinamycin
20	4	5	8	Lovastatin
10	2	3	4	Naphtopyrone
0	0	0	0	Tetronomycin
54	0	35	60	HC-toxin
1	1	1	1	HSAF (Heat stable antifungal factor)

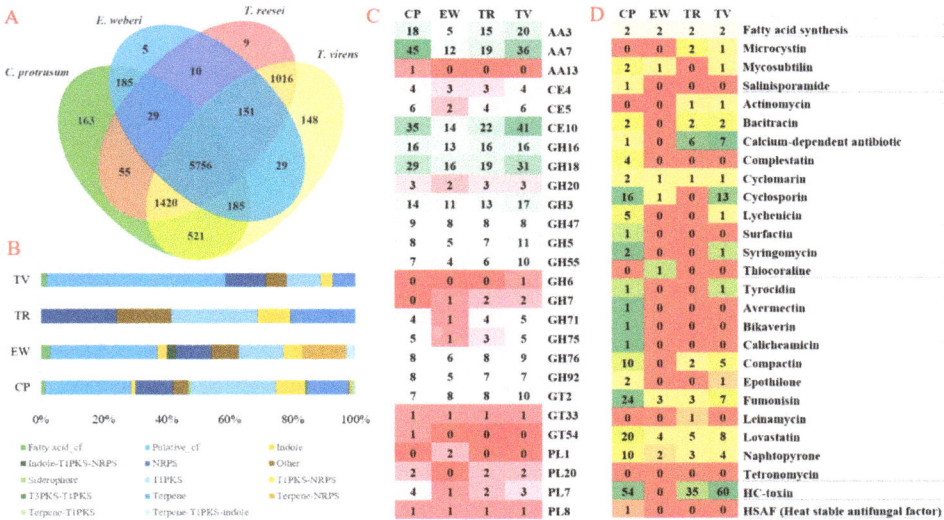

Figure 2. Comparative genomic analysis, carbohydrate-active enzymes (CAZymes) and secondary metabolites (SMs) of *C. protrusum* and three other fungi in the Hypocreaceae family. (**A**) Comparison of the protein-coding genes of *C. protrusum* with those of other Hypocreaceae with different lifestyle *E. weberi* (27.20 Mb, 6870 genes), TR (33.39Mb, 9115 genes) and TV (39.02Mb, 12,406 genes) based on orthology analysis. (**B**) The number of antiSMASH SMs of *C. protrusum* and EW, *T. reesei* (TR), and *T. virens* (TV). (**C**) Abundance of CAZyme modules in *C. protrusum* and EW, TR, and TV. (**D**) The number of NaPDoS SMs of *C. protrusum* and EW, TR, and TV.

3.6. CAZymes in C. protrusum

The genome of *C. protrusum* contains 412 CAZymes with a high diversity of families (Supplementary Tables S5 and S6), including 190 (46.12%) glycoside hydrolases (GH), 77 (18.69%) auxiliary activities (AA), 77 (18.69%) glycosyltransferases (GT), seven (1.70%) polysaccharide lyases (PL), 54 (13.12%) carbohydrate esterase's (CE), and one (0.24%) carbohydrate-binding molecule (CBM). The number of CAZymes possessed by *C. protrusum* is more than that of *E. weberi* (245) and *T. reesei* (366) but is less than that of *T. virens* (484). Most of the differences between *C. protrusum* and *T. virens* can be attributed to the high copy number of GH and CE families. CAZymes involved in oxidative degradation of lignin-based components of the cell wall (10.92%, AA7 = 45) were the most abundant in *C. protrusum* followed by enzymes associated with hydrolysis of carbohydrate and non-carbohydrate substrates (CE10 8.49% (35 genes) and GH18 7.04% (29 genes), respectively). GH and GT enzymes had the largest average sets of genes among the four pathogens studied. The most abundant GH family in *C. protrusum* was GH18 7.04% (29 genes), followed by GH16 with 3.88% (16 genes) and GH3 with 3.40% (14 genes) (Figure 2C). AA13, GH43_14, GH5_11, and GT54 were found to be present in *C. protrusum* but absent in the other three pathogens studied. Other starch degrading enzymes found in all of the species compared were α-amylases of GH13, glucoamylases of GH15, and α-glucosidases of GH31. Therefore, the *C. protrusum* genome contains diverse gene families associated with fungal cell wall synthesis, modification, and degradation.

3.7. Secondary Metabolites in C. protrusum

The genome of *C. protrusum* was enriched with 143 SM gene clusters based on the antiSMASH database using the fungiSMASH option [64]. Only 16 gene clusters had a known function, and the remaining had largely unknown functions and were unique to the fungus (Supplementary Table S10). However, the genomes of *E. weberi*, *T. reesei*, and *T. virens* produced 35, 29, and 106 gene clusters, respectively. Many of the SM gene clusters in *C. protrusum* were grouped into 39 (27.27%) PKSs, 19 (13.29%) terpenes, 17 (11.89%) non-ribosomal peptide synthases (NRPSs), and 13 (9.10%) type 1 PKS-NRPSs (Figure 2B). The number of PKSs in *C. protrusum* was higher than in *Trichoderma* spp. *C. protrusum* has only one NRPS gene cluster (cluster 60), which encodes the apicidin biosynthetic gene cluster siderophore and seven hybrids with type 1PKS. It shares one known gene cluster with *E. weberi*, a fungal antibiotic isoflavipucine, and one with *T. reesei*, an antibiotic LL-Z1272beta, and four known gene clusters with *T. virens*, including nivalenol, vitamin B synthesis biotin, initiate apoptosis cytochalasin, and destruxins. In addition, based on the NaPDoS analysis [65], *C. protrusum*, contains 163 genes that have various functions such as antibiotic, anticancer, anti-inflammatory, immunosuppressant, and phytotoxin functions (Figure 2D). These findings suggest that *C. protrusum* has the potential to produce biologically active compounds.

3.8. Secretory Protein- and Pathogenicity-Related Genes of C. protrusum

The genome of *C. protrusum* was predicted to encode 807 secretory proteins and 428 membrane transport proteins (TCdb database) (Figure 3A, Supplementary Tables S11 and S12). Among these secretory proteins, 378 (46.84%) were predicted to encode cell surface proteins including transmembrane proteins and glycosylphosphatidylinositol GPI-anchored proteins [57], 180 were predicted to encode proteases [59], 180 (43.69%) were predicted to encode CAZymes [63], and 180 were predicted to encode (17.34%) pathogen–host interactions (PHI) [61]. We found that only two secretory proteins were membrane transport proteins. The secretory and membrane transport proteins of *C. protrusum* are similar to the genera *Escovopsis* and *Trichoderma* [66,67]. There are much fewer proteases in *C. protrusum* (Supplementary Table S13) compared to in *Trichoderma* spp. [66], and its common proteases include aspartyl protease, serine carboxypeptidase, lipase, the peptidase family, and subtilase. We also identified 53 (0.48%), 40 (0.36%), and 184 (1.67%) genes that encode for the important family of ATP-binding cassette (ABC) transporters, MFS transporters, and cytochrome P450 (CYP) (Supplementary Table S14), respectively.

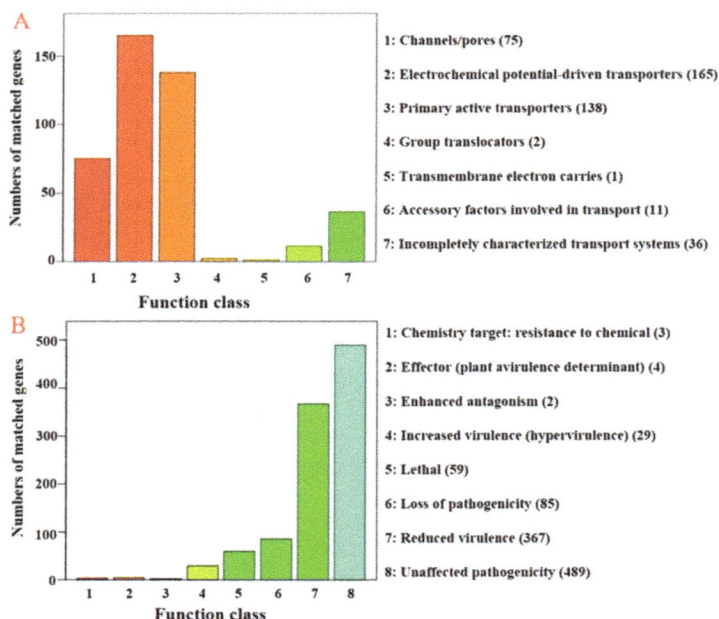

A

1: Channels/pores (75)

2: Electrochemical potential-driven transporters (165)

3: Primary active transporters (138)

4: Group translocators (2)

5: Transmembrane electron carries (1)

6: Accessory factors involved in transport (11)

7: Incompletely characterized transport systems (36)

B

1: Chemistry target: resistance to chemical (3)

2: Effector (plant avirulence determinant) (4)

3: Enhanced antagonism (2)

4: Increased virulence (hypervirulence) (29)

5: Lethal (59)

6: Loss of pathogenicity (85)

7: Reduced virulence (367)

8: Unaffected pathogenicity (489)

Figure 3. Distribution of the number of membrane transport proteins and pathogen–host interaction genes in the *C. protrusum* genome. (**A**) The distribution of membrane transport proteins (TCdb database) in *C. protrusum*; (**B**) the distribution of pathogen–host interaction (PHI) genes in *C. protrusum*. The legends on the right of each graph show the various classifications for each database used.

The complete protein sequences were searched against the PHI [61] and DFVF [62] databases to identify pathogenicity-related genes. We observed a total of 1038 and 453 PHI and DFVF genes, respectively (Figure 3B, Supplementary Tables S15 and S16). Moreover, 47.02% (213) of the DFVF genes were found in the PHI database. The phenotypic classification of PHI genes was classified as follows: chemistry target (0.29%, 3), effectors (plant avirulence determinant) (0.348%, 4), enhanced antagonism (0.19%, 2), increased virulence (hypervirulence) (2.79%, 29), lethal (5.68%, 59), loss of pathogenicity (8.19%, 85), reduced virulence (35.36%, 367), and unaffected pathogenicity (47.11%, 489). For example, we identified the effectors PHI:2118, PHI:2216, PHI:325, and PHI:3123 in the *C. protrusum* genome.

The *C. protrusum* genome encodes 17 (0.16%) fungal G protein coupled receptors (GPCRs); out of these, 12 share homologies with Pth11-like GPCRs (Supplementary Table S17). The number of total GPCRs and Pth11-like GPCRs in the *C. protrusum* genome is much less than in *Trichoderma* spp. from the 65 and 76 putative GPCRs encoded in the *T. atroviride* and *T. virens* genomes, respectively, but much higher than in *E. weberi* [67]. The *C. protrusum* genome contains many other genes, such as hydrophobins, *Thctf1*, *PacC T2* family RNases, *NPP1* (necrosis-inducing protein), GLEYA adhesin domain proteins, killer toxins, and *MD-2-related lipid-recognition* genes, which take part in host pathogenicity, pathogen–host interactions, nutrient acquisition, and adaptation to environmental stress.

Further, we explored the *C. protrusum* genome for mutations in genes that confer antifungal drug resistance by searching the Mycology Antifungal Resistance Database (MARDy) [68]. Out of the 36 antifungal drug resistance gene types in MARDy, nine different genes (*BcSdhB*, *cox10*, *cytb*, *CYP51*, *DHFR*, *DHPS*, *FKS1*, *FUR1*, and *tub2*) were found in *C. protrusum* genome (Supplementary Table S18). There were no mutations in these nine genes, which may indicate a lack of antifungal drug resistance. Hence, there is a need to perform fungicide sensitivity tests to confirm this result.

4. Discussion

C. protrusum is a problematic pathogen that affects mushrooms. Little is currently known about its genomic sequence and structure. In the present study, we performed genome sequencing of *C. protrusum* and a comparative genome analysis to provide insights into its pathogenicity mechanisms. The *C. protrusum* genome size of 39.09 Mb is similar to other reported sizes of Hypocreaceae fungi, which range from 27.14 Mb (*Escovopsis* spp. *AC*) to 40.98 Mb (*T. harzianum* has the largest size in the family so far) [66,67,69]. To date, the genomes of at least 23 members of the family Hypocreaceae have been sequenced, 17 of which are from the genus *Trichoderma* and five from Escovopsis, while *C. protrusum* is the first sequenced genome in the genus *Cladobotryum*. The number of predicted protein-coding genes (11,003) of *C. protrusum* was also consistent with that of other Hypocreaceae fungi, e.g., *T. guizhouensis* (38.33 Mb, 11,255 protein-coding genes) and *T. gamsii* (37.91 Mb, 11,179 protein-coding genes) but lower than those with a similar genome size, e.g., *T. virens* (12,406 protein-coding genes) [66,69]. The number of transposable elements in *C. protrusum* is higher than that reported for members of the genus *Trichoderma*, which lack a significant repetitive DNA component in their genomes [67]. The TE content is variable in different organisms and may be used as a marker to distinguish between clonal populations of *C. protrusum* [70]. The TEs in *C. protrusum* may modify amino acids or contribute to genetic variation, thereby aiding populations to adapt successfully to changes in the environment [71,72]. Previous studies reported that the genome size, structure, and gene content are heavily influenced by natural selection, which is governed by the lifestyle and ecological niche of a species [73].

The genus *Cladobotryum* contains various species with both teleomorph (sexual) and anamorph (asexual) forms [11,74]. The sexual morph of the *Cladobotryum* is classified in a different taxon, which is known as the genus *Hypomyces* [11]. However, there is no known teleomorph for *C. protrusum*.

Mating-type genes control sexual development in fungi [75,76]. We usually use the conserved domains and sequence similarities of MAT genes in fungi to identify the putative mating-type loci [77]. In this study, we found four *MAT1-2* genes in the *C. protrusum genome*, while *MAT1-1* genes were absent. The *MAT1-2* gene encodes the HMG-domain protein, which was highly conserved in comparison with *Metarhizium acridum*, *M. brunneum*, and other ascomycetes [78]. *M. acridum* also lacks the *MAT1-1* idiomorph. Pattemore, et al. [78] suggested that the lack of an observed sexual life cycle may be the result of a loss of gene function, the lack of an opposite mating-type, or merely, the inability to induce a teleomorph under laboratory conditions. Therefore, we suggest that *C. protrusum* is putatively a heterothallic species. Heterothallic fungi need a compatible strain carrying the opposite MAT idiomorph for sex to occur [79]. Therefore, in-depth population sampling is required to confirm if the *MAT1-1* mating-type occurs.

Members of the Hypocreaceae are widely known to have a mycoparasitic lifestyle [17,66,73]. The family Nectriaceae are known to be plant pathogens, while Clavicipitaceae and Ophiocordycipitaceae are insect pathogens as well as parasites of truffle fruiting bodies [73]. Therefore, we performed a phylogenomic analysis using *C. protrusum* and other nine species belonging to the Hypocreaceae, Nectriaceae, Clavicipitaceae, and Ophiocordycipitaceae families. The mimicked recent taxonomic classifications of Hypocreales, which diverged from the MRCA at 332.2 MYA, are in agreement with previous studies [67,80,81]. These results are consistent with recent phylogenetic analyses based on multiple sequence analysis for the family Hypocreaceae [48,67,82]. The results also indicate that *C. protrusum* and *Trichoderma* spp. are distantly related to each other at the family level, which is consistent with their previously assigned phylogenetic placement into different genera based on their morphological and molecular characteristics [66,74]. Mycotrophic behavior is an ancestral lifestyle in the family Hypocreaceae [73]. Different species in the various genera of the family Hypocreaceae have developed different ecological strategies [73,83]; some are aggressive and have a wide host range, like *Trichoderma* species, while others, like *Cladobotryum* spp. and *Escovopsis* spp., are not generally aggressive fungi, but they are highly selective mycoparasites with different species having different host ranges [73].

Compared with other species in the order Hypocreales, *C. protrusum* exhibited a combination of the largest expansion of gene families observed from both *Clonostachys rosea* and *Trichoderma* spp. [17,73]. These expanded gene families encode proteins related to stress, such as transporters, receptors, cell wall proteins, carbohydrate-active enzymes, and SMs (exhibiting high interspecific copy number variation), which also underwent positive selection during the evolution of *C. protrusum*, implying their importance in pathogenicity, adaptation to diverse ecological niches, and host lifestyle [84]. However, the contracted gene families in *C. protrusum* have only one known gene annotation, which is an MFS with high similarity to the one found in *Ophiocordyceps sinensis* and is known to facilitate nutrient transportation [85,86]. Therefore, the expansion of multiple gene families may play a significant role in the pathogenesis and antifungal resistance of *C. protrusum* [87].

Vegetative incompatibility or HET was observed in the *C. protrusum* genome, which is a widespread phenomenon in filamentous fungi [88]. Other proteins associated with HET are the domains of ankyrin, NACHT, and NTPase. There are 83 HET genes in the *C. protrusum* genome, which is much more than the amount observed in other fungi [88–90]. The HET locus inhibits the fusion between two genetically incompatible individuals by forming a fusion cell and undergoing programmed cell death. Ankyrin proteins mediate the protein–protein interactions among HET proteins [88], while NACHT domains are associated with the regulation of apoptosis/programmed cell death in fungi [90]. The lower content of TEs and the lack of a known sexual stage for *C. protrusum* might have influenced the high HET observed in the genome. Therefore, vegetative hyphal fusion controlled by HET genes may be a source of genetic variation, which is vital for the generation of the variability necessary for the adaptation to the environment and to host defense mechanisms [88].

Mycoparasitism depends on a combination of events that include lysis of the cell wall of the host. The number of CAZymes identified in the *C. protrusum* genome was similar to the average reported in other Ascomycetes fungi [91]. Among these CAZymes, the GH18 family is chitinase (-like) proteins associated with the degradation of chitin [92]. The mushroom cell wall is mostly composed of chitin; therefore, chitinolytic enzymes are key factors in mycoparasitic attack [92–94]. Hence, we suggest that the high number of GH18 family members in *C. protrusum* might be mostly used for mycoparasitic attack on mushrooms. The *C. protrusum* genome also includes seven genes encoding GH55 (β-1,3-exoglucanase), and the number of GH55-encoding genes is greater in mycoparasitic *Trichoderma* spp. in comparison with other filamentous fungi (28). Furthermore, most fungi have only one or two chitosanases (GH family 75), while *C. protrusum* have five, similar to the mycoparasitic *T. virens* and *C. rosea* [17,66,73]. Therefore, the results suggest that these cell wall degenerated enzymes play important roles in mycoparasitism for *C. protrusum*.

Comparative analyses of gene content, or paralogous gene number gains or losses, are extensively used to identify genes that are key determinants for ecological niche adaptation [17,66,73,95]. Here, we performed a comparative genome analysis of *C. protrusum* against *E. weberi*, *T. reesei*, and *T. virens* which belong to the same family (Hypocreaceae). *C. protrusum* and *E. weberi* were shown to share the most orthologous gene families. Previous studies showed that SMs produced in fungi are essential for defense, interaction with other organisms, and adaptation to environmental stress [73,95,96]. *Cladobotryum* species have been known to produce SMs (antibiotics) for years [97–100]. These were all higher compared to the other mycoparasites, with the exception of 39 putative gene clusters (cf., putative, a secondary metabolite-related protein that does not fit into any other category) [64], which was higher in *T. virens* (62). The majority of the predicted proteins were similar to proteins present in species of Sordariomycetes, and, in some cases, the best hits were found for species in the Eurotiomycetes taxa.

The PKSs and terpene synthase were shown to belong to SMs, including substances with mycotoxins, conidia, and mycelial pigmentation as well as those with antibiotic, anticancer, anti-cholesterol, anthelmintic, and insecticidal properties, and cholesterol-lowering agents. PKSs and terpene synthase were implicated in the competition and communication between microbes. As mentioned above, the number of PKSs in *C. protrusum* was shown to be higher than in *Trichoderma* spp., and this may be attributed to the higher content of mycotoxin and other genes associated with pigmentation. For instance, the red color of *C. protrusum* mycelia is due to biosynthesis of bikaverin, which was originally found in *Fusarium* species [101] and also acts as an antibiotic against different organisms, such as protozoa, oomycetes, and nematodes. The terpene gene cluster in *C. protrusum* encodes 4,4′-piperazine-2,5-diyldimethyl-bis-phenol, which has high homology to *Aspergillus flavus* and has possible pharmacological properties [102]. Its associated hybrids encode mycotoxins (nivalenol/deoxynivalenol/3-acetyldeoxynivalenol and trichothecene) and phytotoxins (betaenone (A, B and C)). The mycotoxins are similar to those produced by *Aspergillus nidulans*, *Fusarium* spp., and *T. virens*. In addition, we found three velvet genes in the genome. These genes are known to regulate secondary metabolism and mycoparasitism in *Trichoderma* spp. The SM clusters predicted in *C. protrusum* require complete metabolome profiling to confirm the compounds identified.

The secretory proteins are also considered to be important for virulence in fungi, such as proteases, PHI, ABC transporters, CYP, and GPCRs [17,66,93]. Several extracellular proteases including aminopeptidase, metalloproteases, serine carboxypeptidase, lipase, and subtilisin-like proteases were found to play roles in mycoparasitism in *Trichoderma* spp. [17,93,103]. The four known effectors of PHI identified in *C. protrusum* were PHI:2118, PHI:2216, and PHI:325, which were found to cause rice blast disease (*Magnaporthe oryzae*), and PHI:3123, which was found to cause anthracnose (*Colletotrichum orbiculare*) in cucurbits. Two predicted PTH11-encoding genes are also induced in *T. ophioglossoides* during growth on truffle cell wall containing media (34), emphasizing the importance of PTH11-type receptors in Hypocrealean mycoparasites. These related genes are suggested to contribute to the pathogenicity and lifestyle of *C. protrusum*.

Several authors [8,104–106] have reported fungicide resistance in *Cladobotryum* spp. Ma et al. [107] reported that long interspersed element (LINE) transposon of the 14α-demethylase gene (*CYP51*) confer resistance to sterol demethylation inhibitor (DMI) fungicides in *Blumeriella jaapii*. *C. protrusum* genome containing LINE (0.60%). It can therefore be inferred that fungicide resistance in *Cladobotryum* spp. may be a result of mutations in one of the target genes (*BcSdhB*, *cox10*, *cytb*, *CYP51*, *DHFR*, *DHPS*, *FKS1*, *FUR1*, and *tub2*) observed in *C. protrusum*. Currently, measures used to control cobweb disease include strict hygiene, the isolation of infected parts by covering with thick-damp paper to prevent conidial dispersion leading to further outbreaks, and application of fungicides [8,10,11]. This work suggests that it is likely that benzophenone, pyrimidinamines and quinazoline fungicides targeting actin cytoskeleton-regulatory complex protein (PF12761), and (PF12853) NADH-ubiquinone oxidoreductase, respectively, may be valuable for controlling *C. protrusum*. In addition, point mutations in ERG11 gene (*cytochrome P450 lanosterol 14α-demethylase*) which confer azole resistance in *Candida albicans* and *Cryptococcus neoformans* [108,109] were not found in the *C. protrusum* genome. Therefore, any new fungicide targeting the ergosterol biosynthesis ERG4/ERG24 family (PF01222) gene will also be useful for controlling *C. protrusum*.

5. Conclusions

In this study, we sequenced the genome of *C. protrusum*, a pathogen that causes cobweb disease on cultivated edible mushrooms, using the PacBio sequencing platform. The 39.09 Mb genome with 11,003 coding genes is the first sequenced genome for the genus *Cladobotryum*. The analysis confirmed that the fungus belongs to the family Hypocreaceae, and genes from CAZymes, SMs, P450, and PHI all contribute to its mycotrophic lifestyle. Further analysis revealed that *C. protrusum* harbors arrays of genes that potentially produce bioactive SMs and stress response-related proteins that are significant for adaptation to living in hostile environments. Knowledge of the genome sequence will foster a better understanding of the biology of *C. protrusum* and mycoparasitism in general as well as aid in the development of effective disease control strategies to minimize economic losses from cobweb disease in cultivated edible mushrooms.

Supplementary Materials: The following are available online at http://www.mdpi.com/2073-4425/10/2/124/s1, Tables S1–S18.

Author Contributions: Conceptualization, Y.F. and Y.L.; Formal analysis, C.Y., B.A.O., L.S. and Y.F.; Investigation, F.L.S. and Z.L.; Methodology, F.L.S. and Z.L.; Software, C.Y. and L.S.; Supervision, Y.F. and Y.L.; Writing – original draft, F.L.S., Z.L., B.A.O. and Y.F.; Writing – review & editing, F.L.S., Y.F. and Y.L.

Funding: This research was funded by the Special Fund for Agro-scientific Research in the Public Interest (No. 201503137); National Natural Science Foundation of China (No. 31700012); the program of creation and utilization of germplasm of mushroom crop of "111" project (No. D17014); National-level International Joint Research Centre (2017B01011).

Acknowledgments: We thank Prof. Yinbing Bian from the Institute of Applied Mycology, Huazhong Agricultural University, Wuhan, Hubei, China for providing the *C. protrusum* strain used in this study. We are also grateful to Drs. Francis Martin and Stéphane Hacquard, and the 1000 Fungal Genome consortium for giving us access to the *Fusarium solani* unpublished genome data used in the phylogenetic and divergence time tree. The genome sequence data were produced by the US Department of Energy Joint Genome Institute in collaboration with the user community.

Conflicts of Interest: The authors declare no conflict of interest.

References

1. Kertesz, M.A.; Thai, M. Compost bacteria and fungi that influence growth and development of Agaricus bisporus and other commercial mushrooms. *Appl. Microbiol. Biotechnol.* **2018**, *102*, 1639–1650. [CrossRef] [PubMed]
2. Fletcher, J.T.; Hims, M.J.; Hall, R.J. The control of bubble diseases and cobweb disease of mushrooms with prochloraz. *Plant Pathol.* **1983**, *32*, 123–131. [CrossRef]
3. Kim, M.K.; Lee, Y.H.; Cho, K.M.; Lee, J.Y. First Report of Cobweb Disease Caused by Cladobotryum mycophilum on the Edible Mushroom Pleurotus eryngii in Korea. *Plant Dis.* **2012**, *96*, 1374–1374. [CrossRef]

4. Back, C.-G.; Kim, Y.-H.; Jo, W.-S.; Chung, H.; Jung, H.-Y. Cobweb disease on Agaricus bisporus caused by Cladobotryum mycophilum in Korea. *J. Gen. Plant Pathol.* **2010**, *76*, 232–235. [CrossRef]
5. Back, C.-G.; Lee, C.-Y.; Seo, G.-S.; Jung, H.-Y. Characterization of Species of Cladobotryum which Cause Cobweb Disease in Edible Mushrooms Grown in Korea. *Mycobiology* **2012**, *40*, 189–194. [CrossRef] [PubMed]
6. Gea, F.J.; Navarro, M.J.; Suz, L.M. First Report of Cladobotryum mycophilum Causing Cobweb on Cultivated King Oyster Mushroom in Spain. *Plant Dis.* **2011**, *95*, 1030–1030. [CrossRef]
7. Zuo, B.; Lu, B.H.; Liu, X.L.; Wang, Y.; Ma, G.L.; Gao, J. First Report of Cladobotryum mycophilum Causing Cobweb on Ganoderma lucidum Cultivated in Jilin Province, China. *Plant Dis.* **2016**, *100*, 1239–1239. [CrossRef]
8. Gea, F.J.; Carrasco, J.; Suz, L.M.; Navarro, M.J. Characterization and pathogenicity of Cladobotryum mycophilum in Spanish Pleurotus eryngii mushroom crops and its sensitivity to fungicides. *Eur. J. Plant Pathol.* **2017**, *147*, 129–139. [CrossRef]
9. McKay, G.J.; Egan, D.; Morris, E.; Scott, C.; Brown, A.E. Genetic and Morphological Characterization of *Cladobotryum* Species Causing Cobweb Disease of Mushrooms. *Appl. Environ. Microbiol.* **1999**, *65*, 606–610.
10. Grogan, H.M. Fungicide control of mushroom cobweb disease caused by Cladobotryum strains with different benzimidazole resistance profiles. *Pest Manag. Sci.* **2006**, *62*, 153–161. [CrossRef]
11. Tamm, H.; Põldmaa, K. Diversity, host associations, and phylogeography of temperate aurofusarin-producing Hypomyces/Cladobotryum including causal agents of cobweb disease of cultivated mushrooms. *Fungal Biol.* **2013**, *117*, 348–367. [CrossRef] [PubMed]
12. List of Cladobotryum spp. record. Index Fungorum. Available online: http://www.indexfungorum.org/Names/Names.asp (accessed on 8 November 2018).
13. Grigoriev, I.V.; Nikitin, R.; Haridas, S.; Kuo, A.; Ohm, R.; Otillar, R.; Riley, R.; Salamov, A.; Zhao, X.; Korzeniewski, F.; et al. MycoCosm portal: Gearing up for 1000 fungal genomes. *Nucleic Acids Res.* **2014**, *42*, D699–D704. [CrossRef] [PubMed]
14. Buermans, H.P.J.; den Dunnen, J.T. Next generation sequencing technology: Advances and applications. *Biochim. Biophys. Acta (BBA) Mol. Basis Dis.* **2014**, *1842*, 1932–1941. [CrossRef] [PubMed]
15. Rhoads, A.; Au, K.F. PacBio Sequencing and Its Applications. *Genom. Proteom. Bioinform.* **2015**, *13*, 278–289. [CrossRef]
16. Ardui, S.; Ameur, A.; Vermeesch, J.R.; Hestand, M.S. Single molecule real-time (SMRT) sequencing comes of age: Applications and utilities for medical diagnostics. *Nucleic Acids Res.* **2018**, *46*, 2159–2168. [CrossRef] [PubMed]
17. Karlsson, M.; Durling, M.B.; Choi, J.; Kosawang, C.; Lackner, G.; Tzelepis, G.D.; Nygren, K.; Dubey, M.K.; Kamou, N.; Levasseur, A.; et al. Insights on the Evolution of Mycoparasitism from the Genome of Clonostachys rosea. *Genome Biol. Evol.* **2015**, *7*, 465–480. [CrossRef] [PubMed]
18. Wang, G.Z.; Guo, M.P.; Bian, Y.B. First Report of Cladobotryum protrusum causing Cobweb Disease on the Edible Mushroom Coprinus comatus. *Plant Dis.* **2014**, *99*, 287–287. [CrossRef]
19. Parra, G.; Bradnam, K.; Korf, I. CEGMA: A pipeline to accurately annotate core genes in eukaryotic genomes. *Bioinformatics* **2007**, *23*, 1061–1067. [CrossRef]
20. Waterhouse, R.M.; Seppey, M.; Simão, F.A.; Manni, M.; Ioannidis, P.; Klioutchnikov, G.; Kriventseva, E.V.; Zdobnov, E.M. BUSCO Applications from Quality Assessments to Gene Prediction and Phylogenomics. *Mol. Biol. Evol.* **2018**, *35*, 543–548. [CrossRef]
21. Simão, F.A.; Waterhouse, R.M.; Ioannidis, P.; Kriventseva, E.V.; Zdobnov, E.M. BUSCO: Assessing genome assembly and annotation completeness with single-copy orthologs. *Bioinformatics* **2015**, *31*, 3210–3212. [CrossRef]
22. Birney, E.; Clamp, M.; Durbin, R. GeneWise and Genomewise. *Genome Res.* **2004**, *14*, 988–995. [CrossRef] [PubMed]
23. Stanke, M.; Schöffmann, O.; Morgenstern, B.; Waack, S. Gene prediction in eukaryotes with a generalized hidden Markov model that uses hints from external sources. *BMC Bioinform.* **2006**, *7*, 62. [CrossRef] [PubMed]
24. Majoros, W.H.; Pertea, M.; Salzberg, S.L. TigrScan and GlimmerHMM: two open source ab initio eukaryotic gene-finders. *Bioinformatics* **2004**, *20*, 2878–2879. [CrossRef] [PubMed]
25. Burge, C.; Karlin, S. Prediction of complete gene structures in human genomic DNA11Edited by F. E. Cohen. *J. Mol. Biol.* **1997**, *268*, 78–94. [CrossRef] [PubMed]
26. Korf, I. Gene finding in novel genomes. *BMC Bioinform.* **2004**, *5*, 59. [CrossRef] [PubMed]
27. Mackey, A.; Di Giulio, D.; Gilbert, D.; Stajich, J. GLEAN. Available online: https://sourceforge.net/projects/glean-gene/ (accessed on 3 May 2018).

28. Benson, G. Tandem repeats finder: A program to analyze DNA sequences. *Nucleic Acids Res.* **1999**, *27*, 573–580. [CrossRef] [PubMed]
29. Bao, W.; Kojima, K.K.; Kohany, O. Repbase Update, a database of repetitive elements in eukaryotic genomes. *Mobile DNA* **2015**, *6*, 11. [CrossRef]
30. Lowe, T.M.; Eddy, S.R. tRNAscan-SE: A Program for Improved Detection of Transfer RNA Genes in Genomic Sequence. *Nucleic Acids Res.* **1997**, *25*, 955–964. [CrossRef]
31. Lagesen, K.; Hallin, P.; Rødland, E.A.; Staerfeldt, H.-H.; Rognes, T.; Ussery, D.W. RNAmmer: Consistent and rapid annotation of ribosomal RNA genes. *Nucleic Acids Res.* **2007**, *35*, 3100–3108. [CrossRef]
32. Gardner, P.P.; Daub, J.; Tate, J.G.; Nawrocki, E.P.; Kolbe, D.L.; Lindgreen, S.; Wilkinson, A.C.; Finn, R.D.; Griffiths-Jones, S.; Eddy, S.R.; et al. Rfam: Updates to the RNA families database. *Nucleic Acids Res.* **2009**, *37*, D136–D140. [CrossRef]
33. Tatusov, R.L.; Fedorova, N.D.; Jackson, J.D.; Jacobs, A.R.; Kiryutin, B.; Koonin, E.V.; Krylov, D.M.; Mazumder, R.; Mekhedov, S.L.; Nikolskaya, A.N.; et al. The COG database: an updated version includes eukaryotes. *BMC Bioinform.* **2003**, *4*, 41. [CrossRef] [PubMed]
34. Ashburner, M.; Ball, C.A.; Blake, J.A.; Botstein, D.; Butler, H.; Cherry, J.M.; Davis, A.P.; Dolinski, K.; Dwight, S.S.; Eppig, J.T.; et al. Gene ontology: Tool for the unification of biology. The Gene Ontology Consortium. *Nature Genet.* **2000**, *25*, 25–29. [CrossRef] [PubMed]
35. Kanehisa, M.; Furumichi, M.; Tanabe, M.; Sato, Y.; Morishima, K. KEGG: New perspectives on genomes, pathways, diseases and drugs. *Nucleic Acids Res.* **2017**, *45*, D353–D361. [CrossRef] [PubMed]
36. Magrane, M.; Consortium, U. UniProt Knowledgebase: a hub of integrated protein data. *Database* **2011**, *2011*, bar009–bar009. [CrossRef] [PubMed]
37. Bairoch, A.; Apweiler, R. The SWISS-PROT Protein Sequence Data Bank and Its New Supplement TREMBL. *Nucleic Acids Res.* **1996**, *24*, 21–25. [CrossRef] [PubMed]
38. Bairoch, A.; Apweiler, R. The SWISS-PROT protein sequence database and its supplement TrEMBL in 2000. *Nucleic Acids Res.* **2000**, *28*, 45–48. [CrossRef] [PubMed]
39. Mitchell, A.; Chang, H.-Y.; Daugherty, L.; Fraser, M.; Hunter, S.; Lopez, R.; McAnulla, C.; McMenamin, C.; Nuka, G.; Pesseat, S.; et al. The InterPro protein families database: the classification resource after 15 years. *Nucleic Acids Res.* **2015**, *43*, D213–D221. [CrossRef]
40. Liu, W.; Xie, Y.; Ma, J.; Luo, X.; Nie, P.; Zuo, Z.; Lahrmann, U.; Zhao, Q.; Zheng, Y.; Zhao, Y.; et al. IBS: An illustrator for the presentation and visualization of biological sequences. *Bioinformatics (Oxford, UK)* **2015**, *31*, 3359–3361. [CrossRef]
41. Konganti, K.; Wang, G.; Yang, E.; Cai, J.J. SBEToolbox: A Matlab Toolbox for Biological Network Analysis. *Evol. Bioinform.* **2013**, *9*, EBO–S12012. [CrossRef]
42. Li, L.; Stoeckert, C.J., Jr.; Roos, D.S. OrthoMCL: Identification of ortholog groups for eukaryotic genomes. *Genome Res.* **2003**, *13*, 2178–2189. [CrossRef]
43. Edgar, R.C. MUSCLE: Multiple sequence alignment with high accuracy and high throughput. *Nucleic Acids Res.* **2004**, *32*, 1792–1797. [CrossRef] [PubMed]
44. Stamatakis, A. RAxML version 8: A tool for phylogenetic analysis and post-analysis of large phylogenies. *Bioinformatics* **2014**, *30*, 1312–1313. [CrossRef] [PubMed]
45. Darriba, D.; Taboada, G.L.; Doallo, R.; Posada, D. ProtTest 3: Fast selection of best-fit models of protein evolution. *Bioinformatics* **2011**, *27*, 1164–1165. [CrossRef] [PubMed]
46. Yang, Z. PAML 4: Phylogenetic Analysis by Maximum Likelihood. *Mol. Biol. Evol.* **2007**, *24*, 1586–1591. [CrossRef] [PubMed]
47. Sanderson, M.J. r8s: Inferring absolute rates of molecular evolution and divergence times in the absence of a molecular clock. *Bioinformatics* **2003**, *19*, 301–302. [CrossRef] [PubMed]
48. Hedges, S.B.; Marin, J.; Suleski, M.; Paymer, M.; Kumar, S. Tree of life reveals clock-like speciation and diversification. *Mol. Biol. Evol.* **2015**, *32*, 835–845. [CrossRef] [PubMed]
49. De Bie, T.; Cristianini, N.; Demuth, J.P.; Hahn, M.W. CAFE: A computational tool for the study of gene family evolution. *Bioinformatics* **2006**, *22*, 1269–1271. [CrossRef]
50. Ye, X.; Zhong, Z.; Liu, H.; Lin, L.; Guo, M.; Guo, W.; Wang, Z.; Zhang, Q.; Feng, L.; Lu, G.; et al. Whole genome and transcriptome analysis reveal adaptive strategies and pathogenesis of Calonectria pseudoreteaudii to Eucalyptus. *BMC Genom.* **2018**, *19*, 358. [CrossRef]

51. Wang, Y.; Coleman-Derr, D.; Chen, G.; Gu, Y.Q. OrthoVenn: A web server for genome wide comparison and annotation of orthologous clusters across multiple species. *Nucleic Acids Res.* **2015**, *43*, W78–W84. [CrossRef]

52. Emms, D.M.; Kelly, S. OrthoFinder: Solving fundamental biases in whole genome comparisons dramatically improves orthogroup inference accuracy. *Genome Biol.* **2015**, *16*, 157. [CrossRef]

53. Knyaz, C.; Stecher, G.; Li, M.; Kumar, S.; Tamura, K. MEGA X: Molecular Evolutionary Genetics Analysis across Computing Platforms. *Mol. Biol. Evol.* **2018**, *35*, 1547–1549. [CrossRef]

54. Dyrløv Bendtsen, J.; Nielsen, H.; von Heijne, G.; Brunak, S. Improved Prediction of Signal Peptides: SignalP 3.0. *J. Mol. Biol.* **2004**, *340*, 783–795. [CrossRef] [PubMed]

55. Krogh, A.; Larsson, B.; von Heijne, G.; Sonnhammer, E.L.L. Predicting transmembrane protein topology with a hidden markov model: Application to complete genomes11Edited by F. Cohen. *J. Mol. Biol.* **2001**, *305*, 567–580. [CrossRef] [PubMed]

56. Emanuelsson, O.; Nielsen, H.; Brunak, S.; von Heijne, G. Predicting Subcellular Localization of Proteins Based on their N-terminal Amino Acid Sequence. *J. Mol. Biol.* **2000**, *300*, 1005–1016. [CrossRef]

57. Pierleoni, A.; Martelli, P.L.; Casadio, R. PredGPI: A GPI-anchor predictor. *BMC Bioinform.* **2008**, *9*, 392. [CrossRef] [PubMed]

58. Saier, J.M.H.; Reddy, V.S.; Tsu, B.V.; Ahmed, M.S.; Li, C.; Moreno-Hagelsieb, G. The Transporter Classification Database (TCDB): Recent advances. *Nucleic Acids Res.* **2016**, *44*, D372–D379. [CrossRef] [PubMed]

59. Rawlings, N.D.; Barrett, A.J.; Thomas, P.D.; Huang, X.; Bateman, A.; Finn, R.D. The MEROPS database of proteolytic enzymes, their substrates and inhibitors in 2017 and a comparison with peptidases in the PANTHER database. *Nucleic Acids Res.* **2018**, *46*, D624–D632. [CrossRef]

60. Nelson, D.R. The cytochrome p450 homepage. *Hum. Genom.* **2009**, *4*, 59–65. [CrossRef]

61. Urban, M.; Cuzick, A.; Rutherford, K.; Irvine, A.; Pedro, H.; Pant, R.; Sadanadan, V.; Khamari, L.; Billal, S.; Mohanty, S.; et al. PHI-base: A new interface and further additions for the multi-species pathogen–host interactions database. *Nucleic Acids Res.* **2017**, *45*, D604–D610. [CrossRef]

62. Lu, T.; Yao, B.; Zhang, C. DFVF: Database of fungal virulence factors. *Database* **2012**, *2012*, bas032–bas032. [CrossRef]

63. Zhang, H.; Yohe, T.; Huang, L.; Entwistle, S.; Wu, P.; Yang, Z.; Busk, P.K.; Xu, Y.; Yin, Y. dbCAN2: A meta server for automated carbohydrate-active enzyme annotation. *Nucleic Acids Res.* **2018**, *46*, W95–W101. [CrossRef] [PubMed]

64. Blin, K.; Wolf, T.; Chevrette, M.G.; Lu, X.; Schwalen, C.J.; Kautsar, S.A.; de los Santos, E.L.C.; Suarez Duran, H.G.; Kim, H.U.; Nave, M.; et al. AntiSMASH 4.0—improvements in chemistry prediction and gene cluster boundary identification. *Nucleic Acids Res.* **2017**, *45*, W36–W41. [CrossRef] [PubMed]

65. Ziemert, N.; Podell, S.; Penn, K.; Badger, J.H.; Allen, E.; Jensen, P.R. The Natural Product Domain Seeker NaPDoS: A Phylogeny Based Bioinformatic Tool to Classify Secondary Metabolite Gene Diversity. *PLoS ONE* **2012**, *7*, e34064. [CrossRef] [PubMed]

66. Kubicek, C.P.; Herrera-Estrella, A.; Seidl-Seiboth, V.; Martinez, D.A.; Druzhinina, I.S.; Thon, M.; Zeilinger, S.; Casas-Flores, S.; Horwitz, B.A.; Mukherjee, P.K.; et al. Comparative genome sequence analysis underscores mycoparasitism as the ancestral life style of Trichoderma. *Genome Biol.* **2011**, *12*, R40. [CrossRef] [PubMed]

67. De Man, T.J.B.; Stajich, J.E.; Kubicek, C.P.; Teiling, C.; Chenthamara, K.; Atanasova, L.; Druzhinina, I.S.; Levenkova, N.; Birnbaum, S.S.L.; Barribeau, S.M.; et al. Small genome of the fungus *Escovopsis weberi*, a specialized disease agent of ant agriculture. *Proc. Natl. Acad. Sci. USA* **2016**, *113*, 3567–3572. [CrossRef] [PubMed]

68. Nash, A.; Shelton, J.M.G.; Fisher, M.C.; Sewell, T.; Rhodes, J.; Farrer, R.A.; Abdolrasouli, A. MARDy: Mycology Antifungal Resistance Database. *Bioinformatics* **2018**, *34*, 3233–3234. [CrossRef] [PubMed]

69. Druzhinina, I.S.; Chenthamara, K.; Zhang, J.; Atanasova, L.; Yang, D.; Miao, Y.; Rahimi, M.J.; Grujic, M.; Cai, F.; Pourmehdi, S.; et al. Massive lateral transfer of genes encoding plant cell wall-degrading enzymes to the mycoparasitic fungus Trichoderma from its plant-associated hosts. *PLoS Genet.* **2018**, *14*, e1007322. [CrossRef]

70. Amyotte, S.G.; Tan, X.; Pennerman, K.; del Mar Jimenez-Gasco, M.; Klosterman, S.J.; Ma, L.-J.; Dobinson, K.F.; Veronese, P. Transposable elements in phytopathogenic Verticillium spp.: Insights into genome evolution and inter- and intra-specific diversification. *BMC Genom.* **2012**, *13*, 314. [CrossRef]

71. Daboussi, M.J. Fungal transposable elements and genome evolution. *Genetica* **1997**, *100*, 253. [CrossRef]

72. Li, Z.-W.; Hou, X.-H.; Chen, J.-F.; Xu, Y.-C.; Wu, Q.; González, J.; Guo, Y.-L. Transposable Elements Contribute to the Adaptation of Arabidopsis thaliana. *Genome Biol. Evol.* **2018**, *10*, 2140–2150. [CrossRef]

73. Karlsson, M.; Atanasova, L.; Jensen, D.F.; Zeilinger, S. Necrotrophic Mycoparasites and Their Genomes. *Microbiol. Spectr.* **2017**, *5*. [CrossRef]

74. Põldmaa, K. Tropical species of Cladobotryum and Hypomyces producing red pigments. *Stud. Mycol.* **2011**, *68*, 1–34. [CrossRef] [PubMed]

75. Kronstad, J.W.; Staben, C. MATING TYPE IN FILAMENTOUS FUNGI. *Annu. Rev. Genet.* **1997**, *31*, 245–276. [CrossRef] [PubMed]

76. Bennett, R.J.; Turgeon, B.G. Fungal Sex: The Ascomycota. *Microbiol. Spectr.* **2016**, *4*. [CrossRef]

77. Agrawal, Y.; Narwani, T.; Subramanian, S. Genome sequence and comparative analysis of clavicipitaceous insect-pathogenic fungus Aschersonia badia with Metarhizium spp. *BMC Genom.* **2016**, *17*, 367. [CrossRef] [PubMed]

78. Pattemore, J.A.; Hane, J.K.; Williams, A.H.; Wilson, B.A.; Stodart, B.J.; Ash, G.J. The genome sequence of the biocontrol fungus Metarhizium anisopliae and comparative genomics of Metarhizium species. *BMC Genom.* **2014**, *15*, 660. [CrossRef]

79. Seidl, V.; Seibel, C.; Kubicek, C.P.; Schmoll, M. Sexual development in the industrial workhorse *Trichoderma reesei. Proc. Natl. Acad. Sci. USA* **2009**, *106*, 13909–13914. [CrossRef]

80. Sung, G.-H.; Poinar, G.O.; Spatafora, J.W. The oldest fossil evidence of animal parasitism by fungi supports a Cretaceous diversification of fungal–arthropod symbioses. *Mol. Phylogenet. Evol.* **2008**, *49*, 495–502. [CrossRef]

81. Taylor, J.W.; Berbee, M.L. Dating divergences in the Fungal Tree of Life: Review and new analyses. *Mycologia* **2006**, *98*, 838–849. [CrossRef]

82. Maharachchikumbura, S.S.N.; Hyde, K.D.; Jones, E.B.G.; McKenzie, E.H.C.; Bhat, J.D.; Dayarathne, M.C.; Huang, S.-K.; Norphanphoun, C.; Senanayake, I.C.; Perera, R.H.; et al. Families of Sordariomycetes. *Fungal Divers.* **2016**, *79*, 1–317. [CrossRef]

83. Chenthamara, K.; Druzhinina, I.S. 12 Ecological Genomics of Mycotrophic Fungi. In *Environmental and Microbial Relationships*; Druzhinina, I.S., Kubicek, C.P., Eds.; Springer International Publishing: Cham, Switzerland, 2016; pp. 215–246. [CrossRef]

84. Wapinski, I.; Pfeffer, A.; Friedman, N.; Regev, A. Natural history and evolutionary principles of gene duplication in fungi. *Nature* **2007**, *449*, 54. [CrossRef] [PubMed]

85. Liu, Z.-Q.; Lin, S.; Baker, P.J.; Wu, L.-F.; Wang, X.-R.; Wu, H.; Xu, F.; Wang, H.-Y.; Brathwaite, M.E.; Zheng, Y.-G. Transcriptome sequencing and analysis of the entomopathogenic fungus Hirsutella sinensis isolated from Ophiocordyceps sinensis. *BMC Genom.* **2015**, *16*, 106. [CrossRef] [PubMed]

86. Wichadakul, D.; Kobmoo, N.; Ingsriswang, S.; Tangphatsornruang, S.; Chantasingh, D.; Luangsa-ard, J.J.; Eurwilaichitr, L. Insights from the genome of Ophiocordyceps polyrhachis-furcata to pathogenicity and host specificity in insect fungi. *BMC Genom.* **2015**, *16*, 881. [CrossRef]

87. Baroncelli, R.; Amby, D.B.; Zapparata, A.; Sarrocco, S.; Vannacci, G.; Le Floch, G.; Harrison, R.J.; Holub, E.; Sukno, S.A.; Sreenivasaprasad, S.; et al. Gene family expansions and contractions are associated with host range in plant pathogens of the genus Colletotrichum. *BMC Genom.* **2016**, *17*, 555. [CrossRef] [PubMed]

88. Aragona, M.; Minio, A.; Ferrarini, A.; Valente, M.T.; Bagnaresi, P.; Orrù, L.; Tononi, P.; Zamperin, G.; Infantino, A.; Valè, G.; et al. De novo genome assembly of the soil-borne fungus and tomato pathogen Pyrenochaeta lycopersici. *BMC Genom.* **2014**, *15*, 313. [CrossRef] [PubMed]

89. Ohm, R.A.; Feau, N.; Henrissat, B.; Schoch, C.L.; Horwitz, B.A.; Barry, K.W.; Condon, B.J.; Copeland, A.C.; Dhillon, B.; Glaser, F.; et al. Diverse lifestyles and strategies of plant pathogenesis encoded in the genomes of eighteen Dothideomycetes fungi. *PLoS Pathog.* **2012**, *8*, e1003037. [CrossRef] [PubMed]

90. Daskalov, A.; Paoletti, M.; Ness, F.; Saupe, S.J. Genomic clustering and homology between HET-S and the NWD2 STAND protein in various fungal genomes. *PloS ONE* **2012**, *7*, e34854. [CrossRef]

91. Druzhinina, I.S.; Seidl-Seiboth, V.; Herrera-Estrella, A.; Horwitz, B.A.; Kenerley, C.M.; Monte, E.; Mukherjee, P.K.; Zeilinger, S.; Grigoriev, I.V.; Kubicek, C.P. Trichoderma: The genomics of opportunistic success. *Nat. Rev. Microbiol.* **2011**, *9*, 749. [CrossRef]

92. Gruber, S.; Seidl-Seiboth, V. Self versus non-self: Fungal cell wall degradation in Trichoderma. *Microbiology* **2012**, *158*, 26–34. [CrossRef]

93. Xie, B.-B.; Qin, Q.-L.; Shi, M.; Chen, L.-L.; Shu, Y.-L.; Luo, Y.; Wang, X.-W.; Rong, J.-C.; Gong, Z.-T.; Li, D.; et al. Comparative genomics provide insights into evolution of trichoderma nutrition style. *Genome Biol. Evol.* **2014**, *6*, 379–390. [CrossRef]

94. Harman, G.E.; Howell, C.R.; Viterbo, A.; Chet, I.; Lorito, M. Trichoderma species—Opportunistic, avirulent plant symbionts. *Nat. Rev. Microbiol.* **2004**, *2*, 43. [CrossRef] [PubMed]

95. Zeilinger, V.S.a.S. Secondary Metabolites of Mycoparasitic Fungi. *IntechOpen* **2018**. [CrossRef]

96. Calvo, A.M.; Wilson, R.A.; Bok, J.W.; Keller, N.P. Relationship between secondary metabolism and fungal development. *Microbiol. Mol. Biol. Rev. (MMBR)* **2002**, *66*, 447–459. [CrossRef] [PubMed]

97. Sakemi, S.; Bordner, J.; Decosta, D.L.; Dekker, K.A.; Hirai, H.; Inagaki, T.; Kim, Y.-J.; Kojima, N.; Sims, J.C.; Sugie, Y.; et al. CJ-15, 696 and Its Analogs, New Furopyridine Antibiotics from the Fungus *Cladobotryum varium*: Fermentation, Isolation, Structural Elucidation, Biotransformation and Antibacterial Activities. *J. Antibiot.* **2002**, *55*, 6–18. [CrossRef] [PubMed]

98. Bills, G.F.; Platas, G.; Overy, D.P.; Collado, J.; Fillola, A.; Jiménez, M.R.; Martín, J.; del Val, A.G.; Vicente, F.; Tormo, J.R.; et al. Discovery of the parnafungins, antifungal metabolites that inhibit mRNA polyadenylation, from the Fusarium larvarum complex and other Hypocrealean fungi. *Mycologia* **2009**, *101*, 449–472. [CrossRef] [PubMed]

99. Sakamoto, K.; Tsujii, E.; Abe, F.; Nakanishi, T.; Yamashita, M.; Shigematsu, N.; Okuhara, M.; Izumi, S. FR901483, a Novel Immunosuppressant Isolated from *Cladobotryum* sp. No. 11231. *J. Antibiot.* **1996**, *49*, 37–44. [CrossRef] [PubMed]

100. Bastos, C.N.; Neill, S.J.; Horgan, R. A metabolite from Cladobotryum amazonense with antibiotic activity. *Trans. Br. Mycol. Soc.* **1986**, *86*, 571–578. [CrossRef]

101. Wiemann, P.; Willmann, A.; Straeten, M.; Kleigrewe, K.; Beyer, M.; Humpf, H.-U.; Tudzynski, B. Biosynthesis of the red pigment bikaverin in Fusarium fujikuroi: genes, their function and regulation. *Mol. Microbiol.* **2009**, *72*, 931–946. [CrossRef] [PubMed]

102. Moore, G.G.; Mack, B.M.; Beltz, S.B.; Puel, O. Genome sequence of an aflatoxigenic pathogen of Argentinian peanut, Aspergillus arachidicola. *BMC Genom.* **2018**, *19*, 189. [CrossRef] [PubMed]

103. Atanasova, L.; Crom, S.L.; Gruber, S.; Coulpier, F.; Seidl-Seiboth, V.; Kubicek, C.P.; Druzhinina, I.S. Comparative transcriptomics reveals different strategies of Trichodermamycoparasitism. *BMC Genom.* **2013**, *14*, 121. [CrossRef] [PubMed]

104. Grogan, H.M.; Gaze, R.H. Fungicide resistance among Cladobotryum spp.—Causal agents of cobweb disease of the edible mushroom Agaricus bisporus. *Mycol. Res.* **2000**, *104*, 357–364. [CrossRef]

105. McKay, G.J.; Egan, D.; Morris, E.; Brown, A.E. Identification of benzimidazole resistance in Cladobotryum dendroides using a PCR-based method. *Mycol. Res.* **1998**, *102*, 671–676. [CrossRef]

106. Kim, M.K.; Seuk, S.W.; Lee, Y.H.; Kim, H.R.; Cho, K.M. Fungicide Sensitivity and Characterization of Cobweb Disease on a Pleurotus eryngii Mushroom Crop Caused by Cladobotryum mycophilum. *Plant Pathol. J.* **2014**, *30*, 82–89. [CrossRef] [PubMed]

107. Ma, Z.; Proffer, T.J.; Jacobs, J.L.; Sundin, G.W. Overexpression of the 14α-Demethylase Target Gene (*CYP51*) Mediates Fungicide Resistance in *Blumeriella jaapii*. *Appl. Environ. Microbiol.* **2006**, *72*, 2581–2585. [CrossRef] [PubMed]

108. Xiang, M.-J.; Liu, J.-Y.; Ni, P.-H.; Wang, S.; Shi, C.; Wei, B.; Ni, Y.-X.; Ge, H.-L. Erg11 mutations associated with azole resistance in clinical isolates of Candida albicans. *FEMS Yeast Res.* **2013**, *13*, 386–393. [CrossRef]

109. Rodero, L.; Mellado, E.; Rodriguez, A.C.; Salve, A.; Guelfand, L.; Cahn, P.; Cuenca-Estrella, M.; Davel, G.; Rodriguez-Tudela, J.L. G484S amino acid substitution in lanosterol 14-alpha demethylase (ERG11) is related to fluconazole resistance in a recurrent Cryptococcus neoformans clinical isolate. *Antimicrob. Agents Chemother.* **2003**, *47*, 3653–3656. [CrossRef]

![genes logo] *genes*

MDPI

Article

A High-Quality *De novo* Genome Assembly from a Single Mosquito Using PacBio Sequencing

Sarah B. Kingan [1,†], Haynes Heaton [2,†], Juliana Cudini [2], Christine C. Lambert [1],
Primo Baybayan [1], Brendan D. Galvin [1], Richard Durbin [3], Jonas Korlach [1,*] and
Mara K. N. Lawniczak [2,*]

1   Pacific Biosciences, 1305 O'Brien Drive, Menlo Park, CA 94025, USA; skingan@pacb.com (S.B.K.);
    clambert@pacb.com (C.C.L.); pbaybayan@pacb.com (P.B.); bgalvin@pacb.com (B.D.G.)
2   Wellcome Sanger Institute, Wellcome Genome Campus, Hinxton CB10 1SA, UK; whh28@cam.ac.uk (H.H.);
    jc39@sanger.ac.uk (J.C.)
3   Department of Genetics, University of Cambridge, Downing Street, Cambridge CB2 3EH, UK;
    rd109@cam.ac.uk
*   Correspondence: jkorlach@pacb.com (J.K.); mara@sanger.ac.uk (M.K.N.L.)
†   These authors contributed equally to this work.

Received: 18 December 2018; Accepted: 15 January 2019; Published: 18 January 2019

Abstract: A high-quality reference genome is a fundamental resource for functional genetics, comparative genomics, and population genomics, and is increasingly important for conservation biology. PacBio Single Molecule, Real-Time (SMRT) sequencing generates long reads with uniform coverage and high consensus accuracy, making it a powerful technology for *de novo* genome assembly. Improvements in throughput and concomitant reductions in cost have made PacBio an attractive core technology for many large genome initiatives, however, relatively high DNA input requirements (~5 µg for standard library protocol) have placed PacBio out of reach for many projects on small organisms that have lower DNA content, or on projects with limited input DNA for other reasons. Here we present a high-quality *de novo* genome assembly from a single *Anopheles coluzzii* mosquito. A modified SMRTbell library construction protocol without DNA shearing and size selection was used to generate a SMRTbell library from just 100 ng of starting genomic DNA. The sample was run on the Sequel System with chemistry 3.0 and software v6.0, generating, on average, 25 Gb of sequence per SMRT Cell with 20 h movies, followed by diploid *de novo* genome assembly with FALCON-Unzip. The resulting curated assembly had high contiguity (contig N50 3.5 Mb) and completeness (more than 98% of conserved genes were present and full-length). In addition, this single-insect assembly now places 667 (>90%) of formerly unplaced genes into their appropriate chromosomal contexts in the AgamP4 PEST reference. We were also able to resolve maternal and paternal haplotypes for over 1/3 of the genome. By sequencing and assembling material from a single diploid individual, only two haplotypes were present, simplifying the assembly process compared to samples from multiple pooled individuals. The method presented here can be applied to samples with starting DNA amounts as low as 100 ng per 1 Gb genome size. This new low-input approach puts PacBio-based assemblies in reach for small highly heterozygous organisms that comprise much of the diversity of life.

Keywords: low-input DNA; *de novo* genome assembly; long-read SMRT sequencing; mosquito

---

1. Introduction

Exciting efforts to sequence the diversity of life are building momentum [1] but one of many challenges that these efforts face is the small size of most organisms. For example, arthropods, which comprise the most diverse animal phylum, are typically small. Beyond this, while levels of heterozygosity within species vary widely across taxa, intraspecific genetic variation is often highest in small organisms [2]. Over the past two decades, reference genomes for many small organisms have been built through considerable efforts of inbreeding organisms to reduce their heterozygosity levels such that many individuals can be pooled together for DNA extractions. This approach has varied in its success, for example working well for organisms that are easy to inbreed (e.g., many *Drosophila* species [3]), but less well for species that are difficult or impossible to inbreed (e.g., *Anopheles* [4]). Therefore, many efforts to sequence genomes of small organisms have relied primarily on short-read approaches due to the large amounts of DNA required for long-read approaches. For example, the recent release of 28 arthropod genomes as part of the i5K initiative used four different insert size Illumina libraries, resulting in an average contig N50 of 15 kb and scaffold N50 of 1 Mb [5].

Another way to overcome DNA input requirements, while also reducing the number of haplotypes present in a DNA pool, is to limit the number of haplotypes in the pool of individuals by using offspring from a single cross. This is easier than multiple generations of inbreeding, and can be successful. For example, a recent PacBio *Aedes aegypti* assembly used DNA extracted from the offspring of a single cross, thus reducing the maximum number of haplotypes for any given locus to four, thereby improving the assembly process and achieving a contig N50 of 1.3 Mb [6].

However, for an initiative like the Earth BioGenome Project [1] that aims to build high-quality reference genomes for more than a million described species over the next decade, generating broods to reach sufficient levels of high molecular weight DNA for long-read sequencing will be infeasible for the vast majority of organisms. Therefore, new methods that overcome the need to pool organisms are needed to support the creation of reference-quality genomes from wild-caught individuals to increase the diversity of life for which reference genomes can be assembled. Here, we present the first high-quality genome assembled with unamplified DNA from a single individual insect using a new workflow that greatly reduces input DNA requirements.

2. Materials and Methods

2.1. DNA Isolation and Evaluation

High molecular weight DNA was isolated from a single *Anopheles coluzzii* female from the Ngousso colony. This colony was created in 2006 from the broods of approximately 100 wild-caught pure *An. coluzzii* females in Cameroon (pers. comm. Anna Cohuet). Although the colony has been typically held at >100 breeding individuals, given the long time since colonization, there is undoubtedly inbreeding. A single female was ground in 200 μL PBS using a pestle with several up and down strokes (i.e., no twisting), and DNA extraction was carried out using a Qiagen MagAttract HMW kit (PN-67653) following the manufacturer's instructions, with the following modifications: 200 uL 1X PBS was used in lieu of Buffer ATL; PBS was mixed simultaneously with RNAse A, Proteinase K, and Buffer AL prior to tissue homogenisation and incubation; incubation time was shortened to 2 h; solutions were mixed by gently flicking the tube rather than pipetting; and subsequent wash steps were performed for one minute. Any time DNA was transferred, wide-bore tips were used. These modifications were in accordance with recommendations from 10X Genomics HMW protocols that aim to achieve >50 kb molecules. The resulting sample contained ~250 ng of DNA, and we used the FEMTO Pulse (Advanced Analytical, Ankeny, IA, USA) to examine the molecular weight of the resulting DNA. This revealed a relatively sharp band at ~150 kb (Figure S1). The DNA was shipped from the U.K. to California on cold packs, and examined again by running 500 pg on the FEMTO Pulse. While a shift in the molecular weight profile was observed as a result of transport, showing a broader DNA smear with mode of ~40 kb (Figure 1), it was still suitable for library preparation (note that this shifted profile is coincidentally similar to what is observed with the unmodified MagAttract protocol). DNA concentration was determined with a Qubit fluorometer and Qubit dsDNA HS assay kit (Thermo Fisher Scientific, Waltham, MA, USA), and 100 ng from the 250 ng total was used for library preparation.

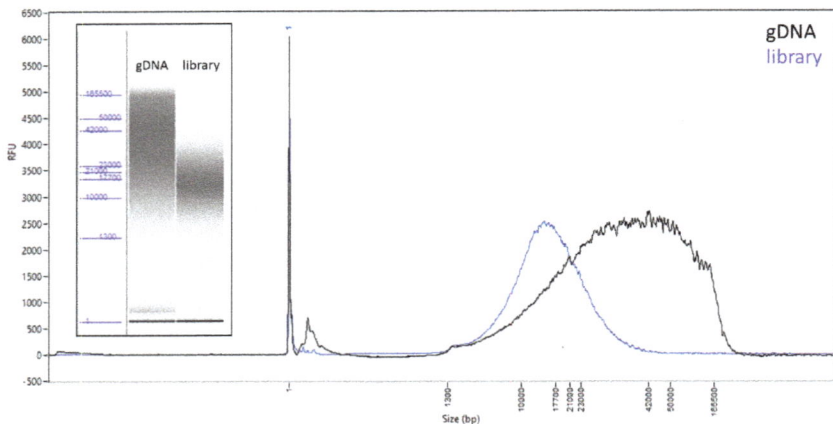

Figure 1. *Anopheles coluzzii* input DNA and resulting library. FEMTO Pulse traces and 'gel' images (inset) of the genomic DNA input (black) and the final library (blue) before sequencing.

2.2. Library Preparation and Sequencing

A SMRTbell library was constructed using an early access version of SMRTbell Express Prep kit v2.0 (Pacific Biosciences, Menlo Park, CA, USA). Because the genomic DNA was already fragmented with the majority of DNA fragments above 20 kb, shearing was not necessary. 100 ng of the genomic DNA was carried into the first enzymatic reaction to remove single-stranded overhangs followed by treatment with repair enzymes to repair any damage that may be present on the DNA backbone. After DNA damage repair, ends of the double stranded fragments were polished and subsequently tailed with an A-overhang. Ligation with T-overhang SMRTbell adapters was performed at 20 °C for 60 min. Following ligation, the SMRTbell library was purified with two AMPure PB bead clean up steps (PacBio, Menlo Park, CA), first with 0.45X followed by 0.80X AMPure. The size and concentration of the final library (Figure 1) were assessed using the FEMTO Pulse and the Qubit Fluorometer and Qubit dsDNA HS reagents Assay kit (Thermo Fisher Scientific, Waltham, MA, USA), respectively.

Sequencing primer v4 and Sequel DNA Polymerase 3.0 were annealed and bound, respectively, to the SMRTbell library. The library was loaded at an on-plate concentration of 5–6 pM using diffusion loading. SMRT sequencing was performed on the Sequel System with Sequel Sequencing Kit 3.0, 1200 min movies with 120 min pre-extension and Software v6.0 (PacBio). A total of 3 SMRT Cells were run.

2.3. Assembly

The genome was assembled using FALCON-Unzip, a diploid assembler that captures haplotype variation in the sample ([7], see Supplementary Materials for software versions and configuration details). A single subread per zero-mode waveguide (ZMW) was used for a total of 12.8 Gb of sequence from three SMRT Cells, or ~48-fold coverage of the ~266 Mb genome. Subreads longer than 4559 bp were designated as "seed reads" and used as template sequences for preassembly/error correction. A total of 8.1 Gb of preassembled reads was generated (~30-fold coverage). After assembly and haplotype separation by FALCON-Unzip, two rounds of polishing were performed to increase the consensus sequence quality of the assembly, aligning the PacBio data to the contigs and computing consensus using the Arrow consensus caller [8]. The first round of polishing was part of the FALCON-Unzip workflow and used a single read per ZMW that was assigned to a haplotype. The second round of polishing was performed in SMRT Link v 6.0.0.43878, concatenating primary contigs and haplotigs into a single reference and aligning all subreads longer than 1000 bp (including multiple subreads from a single sequence read, mean coverage 184-fold) before performing genomic consensus calling. The alignments (BAM files) produced during the two rounds of polishing were used to assess confidence in the contig assembly in regions with rearrangements relative to the AgamP4 PEST assembly for *Anopheles gambiae* (GenBank assembly accession GCA_000005575.2) [8,9]. We referred to the first round of polishing as using "unique subreads" and the second round as using "all subreads".

We explored the performance as a function of the number of SMRT Cells used for the assembly (Table S1), and found that while a single SMRT Cell was insufficient to result in high-quality assembly, data from two SMRT Cells generated a highly contiguous assembly of the correct genome size. We proceeded with the 3-Cell assembly for all subsequent analyses because it gave the best assembly results.

2.4. Curation

The contigs were screened by the Sanger Institute and NCBI to identify contaminants and mitochondrial sequence. Windowmasker was used to mask repeats and the MegaBLAST algorithm was run (with parameter settings: -task megablast -word_size 28 -best_hit_overhang 0.1 -best_hit_score_edge 0.1 -dust yes -evalue 0.0001 -min_raw_gapped_score 100 -penalty −5 -perc_identity 98.0 -soft_masking true -outfmt 7) on the masked genome versus all complete bacterial genomes to find hits with ≥98% homology. In addition, we screened the primary assembly for duplicate haplotypes using Purge Haplotigs [10] with default parameters and coverage thresholds of 20, 150, and 700.

In the process of using PEST to order and orient the PacBio contigs, we found one large potential heterozygous interchromosomal rearrangement between 2L and 3R (Figure S2). Upon further exploration, this was not supported by any subreads mapping across the breakpoint (Figure S2). The putative breakpoints were identified by aligning the PacBio contigs to PEST with minimap2 (asm5 setting), and the start and end position of each aligned subread was determined using bedtools 'bamtobed'. This 4.9 Mb contig had no reads spanning the putative breakpoint when either "unique" or "all subread" alignments were examined and thus we designated this a chimeric misassembly, and split the contig into two.

2.5. Genome Quality Assessment

To assess the completeness of the curated assembly, we searched for conserved, single copy genes using BUSCO (Benchmarking Universal Single-Copy Orthologs) v3.0.2 [11] with the dipteran gene set. In addition, we evaluated assembly completeness against a curated set of genes (AgamP4.10 gene set) from the *An. gambiae* PEST assembly, using a previously described script [12].

To assess the quality of contig assembly and concordance with existing assemblies, the curated primary contigs were aligned to the PEST *Anopheles gambiae* reference genome [8,9] using minimap2 with the "map-pb" settings [13]. For the purpose of comparison, contigs were ordered and oriented according to their median alignment position and orientation on their majority chromosome. A python script with pysam was used in conjunction with ggplot using geom_segments to generate the alignment plots. Large regions (≥250 kb) where assembly contigs did not align to PEST, or where multiple contigs aligned to the same reference region, or where large portions of a single contig aligned discordantly (e.g., to multiple reference chromosomes) were identified and explored manually by visualizing questionable alignments and their breakpoints in the Integrated Genome Browser (IGV, [14]). Confidence in contig assembly was assessed by evaluating subread mapping across putative rearrangement breakpoints as described above. For subread coverage plots, alignments were also made using minimap2 with the "map-pb" setting, and a smoothing filter was applied (mapq 60 filter averaged in 5 kb bins for Figure 3 and Figure S3, and mapq 60 filter averaged in 50 kb bins for Figure 4, respectively) using a custom python script and pysam/numpy. All python scripts referred to above are available [15].

3. Results

3.1. A Modified Protocol Allows for Library Preparation and Sequencing of Samples from as Low as 100 ng of DNA Input

High molecular weight DNA was extracted from a single female mosquito. Given that the genomic DNA had a suitable size range for long-insert PacBio sequencing (Figure 1), the sequencing library preparation protocol was modified to exclude an initial shearing step, which facilitated the use of lower input amounts, as shearing and clean up steps typically lead to loss of DNA material. After following the Express template preparation protocol, the final clean up step was simplified to just two AMPure purification steps to remove unligated adapters and very short DNA fragments, resulting in a final library with a size distribution peak around 15 kb (Figure 1). The library was then sequenced on the Sequel System on 3 SMRT Cells, generating on average 24 Gb of data per SMRT Cell, with average insert lengths of 8.1 kb (insert length N50 ~13 kb, Table S2). The overall library yield was 59%, which would have allowed for the sequencing of at least 8 SMRT Cells, thereby potentially allowing for genome sizes 2–3 times larger than studied here in conjunction with this protocol.

3.2. De novo Assembly Using FALCON-Unzip Allows for a High-Quality Genome from a Single Anopheles coluzzii Mosquito Individual

Using the FALCON-Unzip assembler [7], the resulting primary *de novo* assembly consisted of 372 contigs totaling 266 Mb in length, with half of the assembly in contigs (contig N50) of 3.5 Mb or longer (Table 1). FALCON-Unzip also generated 665 alternate haplotigs, representing regions of sufficient heterozygosity to allow for the separation of the maternal and paternal haplotypes. These additional phased haplotype sequences spanned a total of 78.5 Mb (i.e., 29% of the total genome size was separated into haplotypes), with a contig N50 of 223 kb (Table 1). One contig (#20) was identified as a complete 4.24 Mb bacterial genome, closely related to *Elizabethkingia anophelis*, which is a common gut microbe in *Anopheles* mosquitoes [16]. It was separated from the mosquito assembly and submitted to NCBI separately (see availability of data). We also identified two contigs of mitochondrial origin that each contained multiple copies of the circular chromosome. Full length copies of the mitochondrial chromosome in the higher quality contig differed by only a single base and the consensus sequence was reported as the mitochondrial genome. One of these copies was discarded.

Table 1. Assembly statistics of raw and curated PacBio *Anopheles coluzzii de novo* assembly, compared with the previous Sanger-sequence based assembly for this species from [17] (GCA_000150765.1).

		PacBio Raw	PacBio Curated	Sanger Assembly
Primary contig assembly	Size (Mb)	266	251	224
	No. contigs	372	206	27,063
	Contig N50 (Mb)	3.52	3.47	0.025
Alternate haplotigs	Size (Mb)	78.5	89.2	unresolved
	No. contigs	665	830	N/A
	Contig N50 (Mb)	0.22	0.199	N/A

While FALCON-Unzip resolved haplotypes over ~30% of the genome, 110 genes appeared as duplicated copies in the BUSCO analysis, indicating that highly divergent haplotypes may be assembled as distinct primary contigs as has been observed in other mosquito genome assemblies [6,18]. The presence of duplicated haplotypes can result in erroneously low mapping qualities in resequencing studies and cause problems in downstream scaffolding. Using the "Purge Haplotigs" software [19], we identified 165 primary contigs totalling 10.6 Mb as likely alternate haplotypes, although there remains a possibility that some may be repeats. These contigs were transferred to the alternate haplotig set.

After the above curation steps, including the removal of the bacterial contig, the haplotigs, and the extra copy of the mitochondrial genome, as well as splitting the large chimeric contig, the primary assembly consisted of 206 contigs totaling 251 Mb with contig N50 of 3.47 Mb. Compared to the Sanger sequence based assembly for *An. coluzzii* [17] (AcolM1 Mali-NIH strain assembly AcolM1; GenBank assembly accession GCA_000150765.1), this translated to a reduction in the number of contigs by ~130-fold, as well as an increase in genome contiguity by ~140-fold (Table 1). The PacBio primary assembly was also 12% larger in total size, reflecting additional genomic content that was missing in the previous assembly, corroborated by the conserved gene analysis (see BUSCO analysis results below).

To evaluate genome completeness and sequence accuracy of the assembly, we performed alignment analyses to a set of conserved genes. Using the 'diptera' set of the BUSCO gene collection [11], we observed 98% of the ~2800 genes were complete and >95% occurred as single copies (Table S3). By comparison, the previous assembly had 87.5% complete BUSCO alignments, indicating that a fraction of the genome was missing in that assembly. The percentage of duplicated genes was reduced from 3.9% to 2.4% after curation. Additional analyses are required to distinguish true gene duplication events from incomplete purging of duplicated haplotypes (see discussion below and Figure S3). As an additional evaluation, we aligned to the primary assembly a closely related species gene set (the most recent *An. gambiae* (AgamP4.10) gene set), resulting in 14,972 alignments (99.5%) and an average alignment length of 96.6%, and with >96% of alignments showing no frame shift-inducing indels.

3.3. The New Assembly Shows Improvements in Resolving Genomic Regions

The *An. gambiae* genome, published in 2002, was created using BACs and Sanger sequencing [8]. Further work over the years to order and orient contigs improved this reference [9,20] and to date, AgamP4 [21] remains the highest quality *Anopheles* genome among the 21 that have now been sequenced [4]. However, AgamP4 still has 6302 gaps of Ns in the primary chromosome scaffolds ranging from 20 bases to 36 kb, including 55 gaps of 10 kb that the AGP (A Golden Path) file on Vectorbase annotates as 'contig' endings. The AgamP4 genome was generated from a lab strain known as PEST (Pink Eye STandard) that is long deceased and also was an accidental mixture of two incipient species, previously known as "M" and "S". To address this, the genomes of pure "M" and "S" from new colonies established in Mali were sequenced using only Sanger sequencing [17]. Since then, the "S" form has retained the name *An. gambiae sensu stricto*, and the "M" form has acquired species status and a new name, *An. coluzzii* [22]. It is important to note that while these species show assortative mating, they can hybridise in nature and their hybrids are fully fertile and viable [23]. Given this fact, and the fact that both pure species assemblies remain highly fragmented, we compared our assembly to the best available *An. gambiae* genome (i.e., AgamP4 PEST [21]) to evaluate contiguity and to help order and orient the contigs.

The new PacBio assembly is highly concordant with the AgamP4 PEST reference over the entire genome, allowing the placement of the long PacBio contigs into chromosomal contexts (Figure 2). In addition, the high contiguity of the PacBio contigs allows for the resolution of many gaps in the chromosomal PEST 'contigs'. Note that the only gaps in the PacBio assembly are at contig ends, whereas there are many gaps in PEST that are not annotated as contig breaks so the percent Ns per megabase of PEST is overlaid in the graphs in Figure 2. For example, a single contig from the new PacBio assembly expanded a tandem repeat region on chromosome 2L that in PEST was collapsed, while also filling in many Ns (gaps) in PEST, and also spanning a break between PEST scaffolds set to 10,000 Ns (Figure 3).

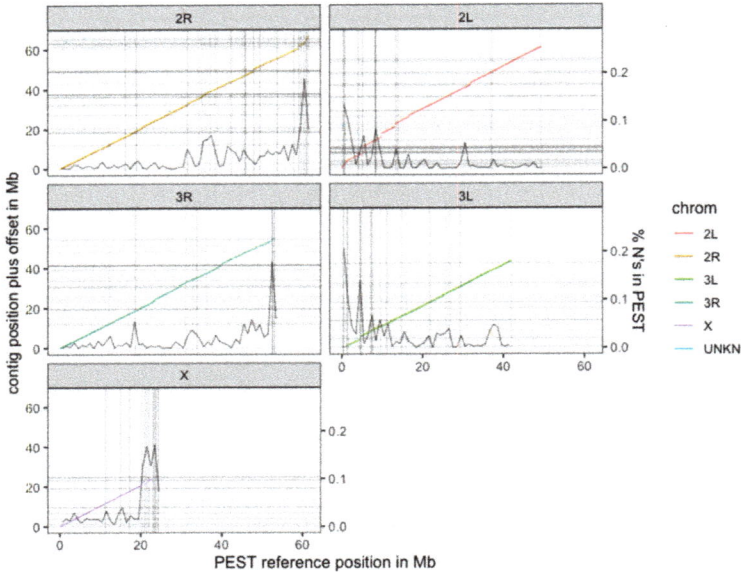

Figure 2. Alignment of the curated PacBio contigs to the AgamP4 PEST reference [21]. Alignments are colored by the primary PEST reference chromosome to which they align but are placed in the panel and *Y* offset to which the contig as a whole aligns best. Contig ends are denoted by horizontal lines in the assembly and vertical lines in PEST. However, there are many Ns in PEST not annotated as contig breaks so the percent Ns per megabase of PEST is overlaid (scale on the right *Y* axis). There are no Ns in the PacBio assembly.

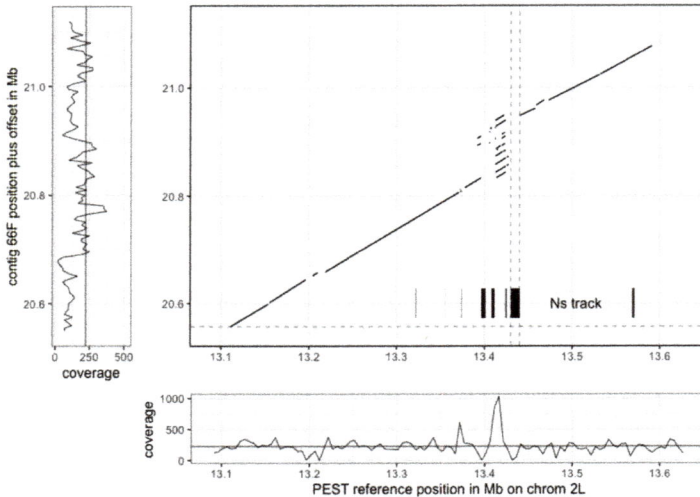

Figure 3. Example of a compressed repeat in PEST that has been expanded by the PacBio assembly. Dotted vertical lines represent a gap in the PEST assembly (10,000 Ns) between scaffolds, which is now spanned by the single PacBio contig. Coverage plot of the PacBio subreads aligned to PEST (bottom) highlights the region where excess coverage indicates a collapsed repeat in PEST, in contrast the coverage of PacBio subreads aligned to the PacBio contig (left) is more uniform.

The PEST annotation also retains a large bin of unplaced contigs (27.3 Mb excluding Ns) designated as the "UNKN" (unknown) chromosome. We compared the alignments of contigs from the PEST chromosomes (X, 2, 3) versus the contigs from the UNKN to the new assembly. Any regions with mapq60 alignments of both UNKN and chromosomal contigs are likely to be haplotigs in the UNKN. In total, we find that 7.27 Mb are haplotigs (i.e., also have PEST chromosomal alignments to the same location in the assembly) and another 10.9 Mb are newly placed sequence that do not overlap with PEST chromosomal alignments. The UNKN bin also contains 737 annotated genes. Remarkably, our single-insect assembly now places 667 (>90%) of these formerly unplaced genes into their appropriate chromosomal contexts (2L:148 genes; 2R:162 genes; 3L: 126 genes; 3R:91 genes; X:140 genes; unplaced:70 genes; details on specific genes can be found in Table S4), which together with their flanking sequence comprise 8.9 Mb of sequence. Altogether, this means that 40% of the UNKN chromosome is now placed in the genome, along with 90% of the genes that were contained within it.

We also identified several potential rearrangements in the 20–22 Mb region of the X chromosome (Figure 4). PEST has contig breaks at the putative breakpoints relative to the assembly, however, given that a single PacBio contig spans the full region and that potential breakpoints relative to PEST are supported by multiple reads, the most likely explanation is an order and orientation issue in PEST, perhaps combined with a potential inversion difference between *An. coluzzii* and the PEST reference. In addition, the contig contains a relatively large region (~380 kb in total) of PacBio sequence corresponding to several pieces in the UNKN section of PEST that can now be assigned to the X chromosome.

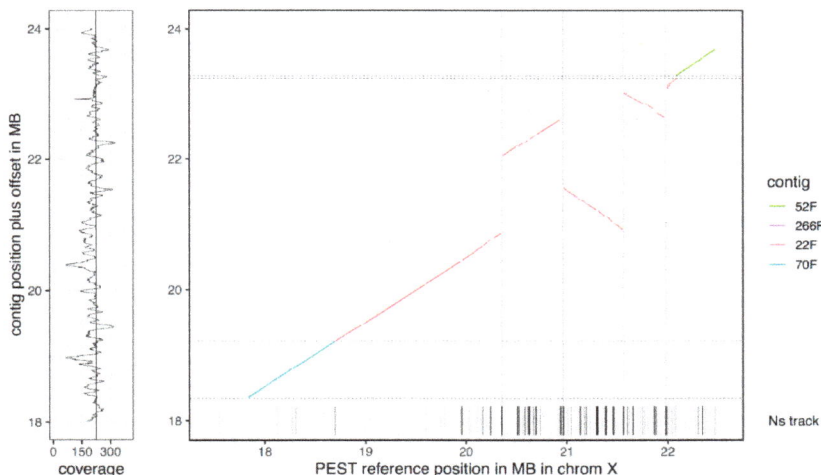

Figure 4. Alignment of X pericentromeric contigs to PEST, highlighting likely order and orientation issues in the PEST assembly that are resolved by a single PacBio contig (22F).

4. Discussion

Long-read PacBio sequencing has been utilized extensively to generate high-quality eukaryote *de novo* genome assemblies, but because of the relatively large DNA input requirements, it has not been used to its full potential for small organisms, requiring time-consuming inbreeding or pooling strategies to generate enough DNA for library preparation and sequencing. Here we present, to our knowledge, the first example of a high-quality *de novo* assembly from a single insect. This assembly, using only one individual and one sequencing technology, exhibits a higher level of contiguity, completeness, accuracy, and degree of haplotype separation than any previous *Anopheles* assembly, demonstrating the impact of long reads on assembly statistics. While the assembly did not achieve independent full chromosomal scale assignment of contigs, its mega-base scale contiguity without gaps immediately provides insights into gene structure and larger-scale genomic architecture, such as promoters, enhancers, repeat elements, large-scale structural variation relative to other species, resolution of tandem repeats (Figure 3), and many other aspects relative to functional and comparative genomics questions.

About a third of the genome for this diploid individual is haplotype-resolved and represented as two separate sequences for the two alleles, thereby providing additional information about the extent and structure of heterozygosity that was not available in previous assemblies, which have been constructed from many pooled individuals. In contrast with approaches requiring multiple individuals, the ability to generate high-quality genomes from single individuals greatly simplifies the assembly process and interpretation, and will allow far clearer lineage and evolutionary conclusions from the sequencing of members of different populations and species. Further, if parental samples are available, the recently developed trio binning assembly approach [24] can be used to further segregate alleles for a full haplotype-resolved assembly of both parental copies of the diploid offspring organism.

The assembly presented here provides an excellent foundation towards generating an improved chromosome-scale reference genome, using the previous PEST reference, scaffolding information from genetic maps, technologies such as Hi-C (e.g., [25]), or alignment of the contigs to closely related species' references. These approaches can also be used to highlight areas of potential improvements to the FALCON-Unzip assembler and to Purge Haplotigs, or other packages used to identify haplotypic contigs. As one example, we noticed in the context of the incomplete haplotype purging described above that some neighboring contig ends exhibited overlaps relative to the PEST reference (Figure S3). The interpretation of such haplotype contig overlaps was corroborated by the observed halving of average sequencing depth over the regions of overlap. These methods could incorporate adjustments to try to account for haplotypic regions in the ends of contigs rather than complete contigs being fully haplotypic.

We noted the importance of the initial DNA size distribution in conjunction with this protocol. Since neither shearing prior to library construction nor size-selection thereafter were employed, the starting high-molecular weight DNA should contain fragments at greater than ~20 kb on average, and without the significant presence of short (smaller than ~5 kb) DNA fragments. Further research into suitable DNA extraction, storage and transportation methodologies is needed to fulfill these requirements for a broader spectrum of different species and environments, in order to allow for the preparation of suitable DNA samples from wild-caught samples originating in sometimes remote areas with limited sample preparation infrastructure.

We anticipate that the new workflow described here will facilitate the sequencing and high-quality assembly of many more species of small organisms, as well as groups of individuals within a species for population-scale analyses, representing an important prerequisite in view of large-scale initiatives such as i5K and the Earth BioGenome Project [1,5]. In addition, other research areas with typically low DNA input regimes could benefit from the described new workflow, e.g., metagenomic community characterizations of small biofilms, DNA isolated from needle biopsy samples, minimization of amplification cycles for targeted or single-cell sequencing applications, and others.

Availability of Data: Raw data and assemblies are deposited in NCBI under BioProject PRJNA508774, and at https://downloads.pacbcloud.com/public/dataset/Mosquito_singleFemale_Assembly/.

Supplementary Materials: The following are available online at http://www.mdpi.com/2073-4425/10/1/62/s1, Figure S1. Femto Pulse evaluation of the Modified MagAttract DNA extraction prior to shipment to California; Figure S2. A chimeric contig between 2L and 3R; Figure S3. Alignment and coverage plot of the PacBio assembly contigs relative to PEST, and magnification of one area of excess coverage; Table S1. Statistics for *Anopheles coluzzii de novo* genome assemblies as a function of the number of SMRT Cells used for the assembly; Table S2. Run statistics for the three Sequel SMRT Cells; Table S3. Analysis of single copy conserved genes using BUSCO v3.0.2 and the diptera gene set; Table S4: Over 90% of genes formerly unassigned to chromosomes (e.g., residing on the UNKN chromosome) have now been placed to chromosomes and contigs; Note S1. Software versions for *de novo* assembly with FALCON-Unzip; Note S2. Configuration file for FALCON assembly; Note S3. Configuration file for FALCON-Unzip.

Author Contributions: M.K.N.L. and J.K. designed the study and led the writing of the manuscript. J.C. performed the DNA extraction, C.C.L., B.D.G., and P.B. developed the library protocol and carried out the sequencing. S.B.K. produced the assembly and assembly evaluations. H.H. performed the assembly evaluation analyses with guidance from R.D. and M.K.N.L. All authors contributed to writing and editing the manuscript.

Funding: M.K.N.L. is supported by an MRC Career Development Award (G1100339) and by the Wellcome Trust (grant 206194/Z/17/Z) to The Wellcome Sanger Institute, J.K. is supported by PacBio funds, and R.D. is supported by Wellcome WT207492. We thank Michelle Smith at the Wellcome Sanger Institute for the Femto Pulse evaluations of the DNA.

Conflicts of Interest: S.B.K., C.C.L., P.B., B.D.G. and J.K. are full-time employees at Pacific Biosciences, a company developing single-molecule sequencing technologies.

References

1. Lewin, H.A.; Robinson, G.E.; Kress, W.J.; Baker, W.J.; Coddington, J.; Crandall, K.A.; Durbin, R.; Edwards, S.V.; Forest, F.; Gilbert, M.T.; et al. Earth BioGenome Project: Sequencing life for the future of life. *Proc. Natl. Acad. Sci. USA* **2018**, *115*, 4325–4333. [CrossRef] [PubMed]
2. Leffler, E.M.; Bullaughey, K.; Matute, D.R.; Meyer, W.K.; Segurel, L.; Venkat, A.; Andolfatto, P.; Przeworski, M. Revisiting an old riddle: What determines genetic diversity levels within species? *PLoS Biol.* **2012**, *10*, e1001388. [CrossRef] [PubMed]
3. *Drosophila* 12 Genomes Consortium; Clark, A.G.; Eisen, M.B.; Smith, D.R.; Bergman, C.M.; Oliver, B. Evolution of genes and genomes on the *Drosophila* phylogeny. *Nature* **2007**, *450*, 203–218. [PubMed]
4. Neafsey, D.E.; Waterhouse, R.M.; Abai, M.R.; Aganezov, S.S.; Alekseyev, M.A.; Allen, J.E.; Amon, J.; Arcà, B.; Arensburger, P.; Artemov, G.; et al. Highly evolvable malaria vectors: The genomes of 16 *Anopheles* mosquitoes. *Science* **2015**, *347*, 1258522. [CrossRef] [PubMed]
5. Thomas, G.W.C.; Dohmen, E.; Hughes, D.S.T.; Murali, S.C.; Poelchau, M.; Glastad, K.; Anstead, C.A.; Ayoub, N.A.; Batterham, P.; Bellair, M.; et al. The Genomic Basis of Arthropod Diversity. *bioRxiv* **2018**, 382945. [CrossRef]
6. Matthews, B.J.; Dudchenko, O.; Kingan, S.B.; Koren, S.; Antoshechkin, I.; Crawford, J.E.; Glassford, W.J.; Herre, M.; Redmond, S.N.; Rose, N.H.; et al. Improved reference genome of Aedes aegypti informs arbovirus vector control. *Nature* **2018**, *563*, 501. [CrossRef] [PubMed]
7. Chin, C.-S.; Peluso, P.; Sedlazeck, F.J.; Nattestad, M.; Concepcion, G.T.; Clum, A.; Dunn, C.; O'Malley, R.; Figueroa-Balderas, R.; Morales-Cruz, A.; et al. Phased diploid genome assembly with single-molecule real-time sequencing. *Nat. Methods* **2016**, *13*, 1050–1054. [CrossRef] [PubMed]
8. Holt, R.A.; Subramanian, G.M.; Halpern, A.; Sutton, G.G.; Charlab, R.; Nusskern, D.R.; Wincker, P.; Clark, A.G.; Ribeiro, J.C.; Wides, R.; et al. The genome sequence of the malaria mosquito *Anopheles gambiae*. *Science* **2002**, *298*, 129–149. [CrossRef]
9. Sharakhova, M.V.; Hammond, M.P.; Lobo, N.F.; Krzywinski, J.; Unger, M.F.; Hillenmeyer, M.E.; Bruggner, R.V.; Birney, E.; Collins, F.H. Update of the *Anopheles gambiae* PEST genome assembly. *Genome Biol.* **2007**, *8*, R5. [CrossRef]
10. Roach, M.J.; Schmidt, S.A.; Borneman, A.R. Purge Haplotigs: Allelic contig reassignment for third-gen diploid genome assemblies. *BMC Bioinform.* **2018**, *19*, 460. [CrossRef]
11. Waterhouse, R.M.; Seppey, M.; Simão, F.A.; Manni, M.; Ioannidis, P.; Klioutchnikov, G.; Kriventseva, E.V.; Zdobnov, E.M. BUSCO applications from quality assessments to gene prediction and phylogenomics. *Mol. Biol. Evol.* **2017**, *35*, 543–548. [CrossRef] [PubMed]
12. Korlach, J.; Gedman, G.; Kingan, S.B.; Chin, C.-S.; Howard, J.T.; Audet, J.-N.; Cantin, L.; Jarvis, E.D. *De novo* PacBio long-read and phased avian genome assemblies correct and add to reference genes generated with intermediate and short reads. *GigaScience* **2017**, *6*, 1–16. [CrossRef] [PubMed]
13. Li, H. Minimap2: Pairwise alignment for nucleotide sequences. *Bioinformatics* **2018**, *1*, 7. [CrossRef] [PubMed]
14. Thorvaldsdóttir, H.; Robinson, J.T.; Mesirov, J.P. Integrative Genomics Viewer (IGV): High-performance genomics data visualization and exploration. *Brief. Bioinform.* **2013**, *14*, 178–192. [CrossRef] [PubMed]
15. Python Assembly Comparison Scripts [Internet]. Available online: https://github.com/wheaton5/assembly_comparison_scripts.

16. Kukutla, P.; Lindberg, B.G.; Pei, D.; Rayl, M.; Yu, W.; Steritz, M.; Faye, I.; Xu, J. Insights from the genome annotation of Elizabethkingia anophelis from the malaria vector *Anopheles gambiae*. *PLoS ONE* **2014**, *9*, e97715. [CrossRef] [PubMed]

17. Lawniczak, M.K.; Emrich, S.J.; Holloway, A.K.; Regier, A.P.; Olson, M.; White, B.; Redmond, S.; Fulton, L.; Appelbaum, E.; Godfrey, J.; et al. Widespread divergence between incipient *Anopheles gambiae* species revealed by whole genome sequences. *Science* **2010**, *330*, 512–514. [CrossRef]

18. Ghurye, J.; Koren, S.; Small, S.T.; Redmond, S.; Howell, P.; Phillippy, A.M.; Besansky, N.J. A chromosome-scale assembly of the major African malaria vector *Anopheles funestus*. *bioRxiv* **2018**, 492777. [CrossRef]

19. Roach, M.J.; Schmidt, S.A.; Borneman, A.R. Purge Haplotigs: Synteny Reduction for Third-gen Diploid Genome Assemblies. *bioRxiv* **2018**. [CrossRef]

20. Sharakhova, M.V.; George, P.; Brusentsova, I.V.; Leman, S.C.; Bailey, J.A.; Smith, C.D.; Sharakhov, I.V. Genome mapping and characterization of the *Anopheles gambiae* heterochromatin. *BMC Genom.* **2010**, *11*, 459. [CrossRef]

21. AgamP4 | VectorBase. Available online: https://www.vectorbase.org/organisms/anopheles-gambiae/pest/agamp4 (accessed on 7 August 2018).

22. Coetzee, M.; Hunt, R.H.; Wilkerson, R.; Torre, A.D.; Coulibaly, M.B.; Besansky, N.J. *Anopheles coluzzii* and *Anopheles amharicus*, new members of the *Anopheles gambiae* complex. *Zootaxa* **2013**, *3619*, 246–274. [CrossRef]

23. Aboagye-Antwi, F.; Alhafez, N.; Weedall, G.D.; Brothwood, J.; Kandola, S.; Paton, D.; Fofana, A.; Olohan, L.; Betancourth, M.P.; Ekechukwu, N.E.; et al. Experimental Swap of *Anopheles gambiae*'s Assortative Mating Preferences Demonstrates Key Role of X-Chromosome Divergence Island in Incipient Sympatric Speciation. *PLoS Genet.* **2015**, *11*, e1005141. [CrossRef] [PubMed]

24. Koren, S.; Rhie, A.; Walenz, B.P.; Dilthey, A.T.; Bickhart, D.M.; Kingan, S.B.; Hiendleder, S.; Williams, J.L.; Smith, T.P.; Phillippy, A.M. *De novo* assembly of haplotype-resolved genomes with trio binning. *Nat. Biotechnol.* **2018**, *36*, 1174. [CrossRef] [PubMed]

25. Kronenberg, Z.N.; Hall, R.J.; Hiendleder, S.; Smith, T.P.; Sullivan, S.T.; Williams, J.L.; Kingan, S.B. FALCON-Phase: Integrating PacBio and Hi-C data for phased diploid genomes. *bioRxiv* **2018**, 327064. [CrossRef]

genes

MDPI

Article

Genome Assembly and Annotation of the *Trichoplusia ni* Tni-FNL Insect Cell Line Enabled by Long-Read Technologies

Keyur Talsania [1], Monika Mehta [2], Castle Raley [2], Yuliya Kriga [2], Sujatha Gowda [2], Carissa Grose [3], Matthew Drew [3], Veronica Roberts [3], Kwong Tai Cheng [3], Sandra Burkett [4], Steffen Oeser [5], Robert Stephens [3], Daniel Soppet [2], Xiongfeng Chen [1], Parimal Kumar [2], Oksana German [2], Tatyana Smirnova [2], Christopher Hautman [2], Jyoti Shetty [2], Bao Tran [2], Yongmei Zhao [1,*] and Dominic Esposito [3,*]

[1] Advanced Biomedical Computational Science, Frederick National Laboratory for Cancer Research sponsored by the National Cancer Institute, Frederick, MD 21701, USA; keyur.talsania@nih.gov (K.T.); xiongfong.chen2@nih.gov (X.C.)
[2] Cancer Research Technology Program, Frederick National Laboratory for Cancer Research Sponsored by the National Cancer Institute, Frederick, MD 21701, USA; monika.mehta@nih.gov (M.M.); castleraley@gwu.edu (C.R.); krigay@mail.nih.gov (Y.K.); sujatha.gowda2@nih.gov (S.G.); soppetdr@mail.nih.gov (D.S.); parimal.kumar@nih.gov (P.K.); oksana.german@nih.gov (O.G.); tatyana.smirnova@nih.gov (T.S.); christopher.hautman@nih.gov (C.H.); jyoti.shetty@nih.gov (J.S.); bao.tran@nih.gov (B.T.)
[3] NCI RAS Initiative, Frederick National Laboratory for Cancer Research Sponsored by the National Cancer Institute, Frederick, MD 21701, USA; carissa.grose@nih.gov (C.G.); matt.drew@nih.gov (M.D.); veronica.roberts@nih.gov (V.R.); oscar.cheng@nih.gov (K.T.C.); stephensr@mail.nih.gov (R.S.)
[4] Comparative Molecular Cytogenetics Core Facility, Frederick National Laboratory for Cancer Research sponsored by the National Cancer Institute, Frederick, MD 21701, USA; sandra.burkett@nih.gov
[5] Bionano Genomics, San Diego, CA 92121, USA; soeser@bionanogenomics.com
* Correspondence: yongmei.zhao@nih.gov (Y.Z.); dominic.esposito@nih.gov (D.E.); Tel.: +1-301-360-3455 (Y.Z.); +1-301-846-7376 (D.E.)

Received: 17 December 2018; Accepted: 14 January 2019; Published: 23 January 2019

Abstract: Background: *Trichoplusia ni* derived cell lines are commonly used to enable recombinant protein expression via baculovirus infection to generate materials approved for clinical use and in clinical trials. In order to develop systems biology and genome engineering tools to improve protein expression in this host, we performed de novo genome assembly of the *Trichoplusia ni*-derived cell line Tni-FNL. Methods: By integration of PacBio single-molecule sequencing, Bionano optical mapping, and 10X Genomics linked-reads data, we have produced a draft genome assembly of Tni-FNL. Results: Our assembly contains 280 scaffolds, with a N50 scaffold size of 2.3 Mb and a total length of 359 Mb. Annotation of the Tni-FNL genome resulted in 14,101 predicted genes and 93.2% of the predicted proteome contained recognizable protein domains. Ortholog searches within the superorder *Holometabola* provided further evidence of high accuracy and completeness of the Tni-FNL genome assembly. Conclusions: This first draft Tni-FNL genome assembly was enabled by complementary long-read technologies and represents a high-quality, well-annotated genome that provides novel insight into the complexity of this insect cell line and can serve as a reference for future large-scale genome engineering work in this and other similar recombinant protein production hosts.

Keywords: de novo assembly; PacBio single molecule real-time sequencing; *Tricoplusia ni*; insect genome; next generation sequencing; optical mapping

1. Introduction

Cell lines derived from *Trichoplusia ni*, the cabbage looper moth, have been used for many years to produce recombinant proteins by means of the baculovirus expression vector system (BEVS). While cell lines from other lepidopteran hosts such as *Spodoptera frugiperda* have commonly been used for production of baculoviruses, *Trichoplusia* cell lines have been shown in several cases to out-perform these cell lines for production yield and protein quality, particularly with regard to secreted proteins [1]. In the past decade, insect cell protein production has emerged as a viable alternative to bacterial and mammalian cells for the production of therapeutically relevant proteins, with several vaccine products generated in baculovirus-infected insect cells having been approved by regulatory agencies [2,3]. Therefore, a comprehensive systems biology approach to improving protein production in these cell lines would be of significant benefit to their potential utility as protein production hosts. However, to date, there are only a few complete genome sequences of lepidopteran hosts. One of them is the silkworm *Bombyx mori*, that has been published [4], while an incomplete draft genome of *Spodoptera frugiperda* (the host from which Sf9 and Sf21 lines were derived) is the only sequence available for the more commonly used protein production hosts [5]. The transcriptome [6] of the *Trichoplusia ni*-derived cell line, Tnms42, and RNA-seq data [7] from the High Five cell line (BTI-Tn-5B1-4), have been published, but these data are not useful for large-scale genome engineering due to a lack of non-coding genomic DNA information. In addition, transcriptome data are inherently biased towards genes with high transcription levels, and likely lack coverage of significant regions of the coding genome.

Tni-FNL is a cell line derived by adaptation of BTI-Tn-5B1-4 cells, originally isolated from *Trichoplusia ni* egg cells in the Wood laboratory at Cornell [8]. While the original cell line was an adherent cell line that grew in the presence of serum, Tni-FNL was selected for suspension growth to optimize its utility for protein production and for the ability to grow in the absence of serum. The Tni-FNL cell line has been shown to routinely produce higher levels of protein than Sf9 or Sf21 cells, and in some cases to surpass the levels produced in the more commonly used and commercially available *Trichoplusia ni* cell line, High Five. High Five cells were derived from the same parent line as Tni-FNL, suggesting that the specific process used for adaptation likely effected changes in the cell line, which resulted in this improvement in protein production. For these reasons, we decided to elucidate the complete genome sequence of the Tni-FNL cell line. This will benefit systems biology approaches to create improved cell lines that can support higher levels of protein expression and potentially improve the quality and lower the cost of therapeutic protein production.

Next-generation sequencing technologies have long been used in the genome assembly of many animal and plant genomes. However, the short-reads they produce have difficulty spanning repetitive regions commonly found in many genomes, and therefore, generate draft genomes consisting of many gaps with potential mis-assemblies and collapsed contigs. Recent advances in sequencing technologies, especially in single-molecule sequencing [9], have resulted in the ability to sequence reads that are longer than most of the common repeats in both microbial and vertebrate genomes, leading to the generation of highly contiguous assemblies. Combining PacBio single-molecule sequencing [9] with complementary technologies such as Illumina short reads, Bionano optical mapping [10], and 10X Genomics (Pleasanton, CA, USA) linked reads [11] has become the recommended strategy for optimal genome assembly [12]. Here we report that by applying the new technologies and assembly strategies, we have generated the first draft genome assembly of Tni-FNL cell line, which was derived from *Trichoplusia ni* cells. Comparative analysis between our draft genome of the Tni-FNL (*Trichoplusia ni*) genome with other closely related species, as well as the recently published Hi5 germ cell genome assembly [13], provided further evidence of high accuracy and completeness of our Tni-FNL cell line genome assembly.

2. Materials and Methods

2.1. Cell Culture Conditions

Tni-FNL cells were cultured under shaking conditions (125 rpm, 2-inch throw) in 500 mL shaker flasks at 27 °C in Gibco Sf-900 III SFM media (Thermo Fisher Scientific, Waltham, MA, USA).

2.2. PacBio Library Preparation and Sequencing

High-molecular-weight genomic DNA (20–150 kb) was extracted from the cultured Tni-FNL cell line using the Genomic-tip 20/G kit (Qiagen, Hilden, Germany). For PacBio library preparation, approximately 15 μg of genomic DNA were sheared to an average size of 20 kb using a g-TUBE™ (Covaris®, Woburn, MA, USA). All sizing and quantitation measurements were performed using the genomic kit for the TapeStation 2200 (Agilent Technologies, Santa Clara, CA, USA). Purity was assessed by calculating the ratio of absorbance at 260 nm to absorbance at 280 nm as measured on a NanoDrop™ spectrophotometer (Thermo Fisher Scientific, Waltham, MA, USA) and was determined to be suitable. Following PacBio's standard 20 kb library preparation protocol 100-286-000-05, the final library was size selected using a dye-free 0.75% agarose cassette on a BluePippin (Sage Science, Beverly, MA, USA) with a lower cutoff of 10 kb. 16 SMRT® cells were sequenced on the PacBio RS II (Pacific Biosciences, Menlo Park, CA, USA) using P6/C4 chemistry, 0.15 nM MagBead loading concentration, and 360 min movie lengths. Additionally, 5 μg of genomic DNA from the same sample were sheared to an average size of 20 kb using a g-TUBE (Covaris, Woburn, MA, USA), which was used as input to create a library using the Accel-NGS® XL Library Kit for Pacific Biosciences® (Swift Biosciences™, Ann Arbor, MI, USA). The final library was size selected using a dye-free 0.75% agarose cassette on a BluePippin (Sage Science, Beverly, MA, USA) with a lower cutoff of 15 kb. 2 SMRT® cells were sequenced on the PacBio RS II (Pacific Biosciences, Menlo Park, CA, USA) using P6/C4 chemistry, 0.15 nM MagBead loading concentration, and 360 min movie lengths. Additionally, approximately 15 μg of genomic DNA were sheared to an average size of 20 kb and were prepared for sequencing on the PacBio Sequel System, using a size selection with a 15 kb lower cutoff on the BluePippin. Two Sequel 1M SMRT® Cells were sequenced using Sequel Polymerase 2.0 and Sequel Sequencing Kit 2.0 (Pacific Biosciences, Menlo Park, CA, USA).

2.3. Bionano Optical Mapping

Optical mapping was performed using the Irys optical mapping technology from Bionano Genomics (San Diego, CA, USA). The sample was prepared as per the IrysPrep Plug Lysis protocol 30026 Rev D and Labeling-NLRS protocol 30024 Rev J. Two million cells from the Tni-FNL cell line were embedded in an agarose plug for extraction of ultra-high-molecular-weight genomic DNA (100–2000 kb). Briefly, the cells were washed with phosphate buffered saline (PBS), the cell suspension was mixed thoroughly with 2% Agarose, and then set into cold plug molds for 15 min. Plugs were treated overnight with Proteinase K at 50 °C, followed by RNase A digestion at 37 °C for 1 h. After washing the plugs with wash buffer and TE, DNA were recovered by incubating the molten plug with Agarase for 45 min at 43 °C. The DNA were further cleaned by Drop Dialysis using a 0.1 μm dialysis membrane set on top of TE in a petri dish. The DNA were dispensed on top of the membrane and dialyzed for 45 min. Homogenization of the DNA was achieved by overnight incubation at room temperature. 600 ng of purified high molecular weight DNA were nicked using 80 Units of the nicking endonuclease Nb.BssSI (New England Biolabs, Ipswich, MA, USA) for 2 h at 37 °C. Fluorescently tagged nucleotides were then incorporated at the nicked sites by Taq DNA polymerase during the labeling reaction at 72 °C for 60 min. This was followed by repair in the presence of polymerase and Taq DNA ligase for 30 min at 37 °C. After counterstaining the DNA backbone with the YOYO-1 dye, the final sample was quantitated again and 9 μL were loaded into each flowcell of an IrysChip. The labeled DNA molecules were linearized in the nanochannels on the chip and imaged by the Irys instrument (Bionano Genomics, San Diego, CA, USA). Both flowcells were run per the Modified Base

Recipe for 30 cycles, with the DNA concentration time of 200 s. After the first run, pillar cleaning of the chip was performed, and the chip was imaged again for an additional 30 cycles. This over-cycling was performed three additional times for both flowcells of the IrysChip (30 cycles in each run) to acquire additional data.

2.4. 10X Genomics Linked Reads Sequencing

High-molecular weight DNA from the Hi5 cells (extracted using the Bionano Plug Lysis protocol) was also used to make 10X libraries, as per the Chromium Genome library preparation protocol from 10X Genomics (Pleasanton, CA, USA). In brief, 0.9 ng/µL DNA were used for GEM generation in the Chromium Controller machine (10X Genomics, Pleasanton, CA, USA). The long DNA molecules were partitioned along with oligo-coated Gel Beads that provide a 16 bp 10X barcode, an Illumina R1 sequence, and a 6 bp random primer sequence. Isothermal incubation of the GEMs at 30 °C for 3 h, followed by 65 °C for 10 min produced barcoded fragments. These fragments were recovered from the GEMs and cleaned up for subsequent library preparation steps that included end repair, A-tailing and adapter ligation per the manufacturer's recommendations. Eight cycles of amplification during the sample index PCR provided enough yield of the indexed library. The library was quantitated by qPCR and sequenced on NextSeq (High output kit) (Illumina, San Diego, CA, USA) with 2 × 150 paired-end reads.

2.5. Transcriptome Sequencing of Tni-FNL Cell Line

Total RNA was extracted from Tni-FNL cells using the NEB Monarch Total RNA Miniprep kit (New England Biolabs, Ipswich, MA, USA) as per the manufacturer's instructions. Briefly, a frozen pellet of approximately 20 million cells was thawed and resuspended in RNA lysis buffer. Genomic DNA was removed by binding to a gDNA removal column, followed by purification of RNA by binding to the RNA purification column. On-column DNase I treatment was performed for removal of residual gDNA. RNA was eluted in nuclease-free water and RNA integrity was assessed using the RNA 6000 Nano Kit on an Agilent 2100 Bioanalyzer (Agilent Technologies, Santa Clara, CA, USA). Approximately 20 µg RNA was obtained and aliquoted before storage and further use.

Short-read sequencing library was prepared from the Tni-FNL total RNA using the NEBNext Ultra II Directional RNA Library Prep kit (New England Biolabs, Ipswich, MA, USA) as per the manufacturer's instructions. Briefly, 1 µg and 500 ng total RNA was subjected to rRNA depletion by rRNA probe hybridization and RNase H digestion. The excess probes were removed by DNase I digestion and the RNA purified using RNAClean XP beads (Beckman Coulter, Brea, CA, USA). RNA was fragmented at 94 °C for 5 min to generate a large insert library. Accordingly, longer incubation time (50 min at 42 °C) was used during first strand cDNA synthesis. Purified double-stranded cDNA was subjected to end prep and adaptor ligation as per the protocol. After purification and selection for larger size fragments, the adaptor ligated DNA was enriched using 9 cycles of PCR amplification and purified using SPRIselect beads (Beckman Coulter, Brea, CA, USA). Library quality assessment on Agilent 2100 Bioanalyzer (Agilent Technologies, Santa Clara, CA, USA) revealed the average library sizes to be around 430 bp.

Total RNA libraries generated from 1 µg and 500 ng Tni-FNL RNA were pooled and quantified by qPCR. Paired-end sequencing (150 bp reads) was performed on the Illumina (San Diego, CA, USA) MiSeq platform using v2 sequencing chemistry.

2.6. Propidium Iodide (PI) Staining for Ploidy Determination

Tni-FNL and Sf9 cells in exponential growth phase were harvested at a cell concentration of approximately 1×10^6 cells/mL and centrifuged at 1050 rpm for 5 min after which the supernatant was discarded. The cell pellet was suspended in 20 mL of cold (−20 °C) 70% ethanol for fixation and the samples were stored at −20 °C. On the day of flow cytometry analysis, cell samples were centrifuged at 1050 rpm for 5 min to remove the ethanol fixative. The supernatant was discarded and the cells were washed two times with 10 mL phosphate buffered saline (PBS). The sample was centrifuged again at 1050 rpm for 5 min, the supernatant was discarded, and the pellet was suspended in 1 mL of RNase solution (250 µg/mL; Sigma, St. Louis, MO, USA) for 20 min at 37 °C. A 50 µL aliquot of propidium iodide (PI; 50 µg/mL) was added to each sample, mixed, and incubated at room temperature for 5 min, before analysis by flow cytometer.

2.7. Flow Cytometry Analysis

Flow cytometric analysis was performed using the LSRFortessa (BD Biosciences, San Jose, CA, USA) SSC-A vs FSC-A with a gate for cell population. Single cells were selected for analysis by using the distribution of propidium iodide-W against propidium iodide-A to discriminate doublets and debris. The propidium iodide-A voltage was adjusted to set the mean of the singlet peak of the Sf9 cell (reference cell) G0/G1 population at 50,000 in the histogram. The data were collected using FACSDiva software version 8.0 (BD Biosciences, San Jose, CA, USA) and analyzed by FlowJo software version 10.2 (FlowJo, Ashland, OR, USA). The DNA index was calculated as the ratio of the mean fluorescence intensity (MFI) of the Tni-FNL cell G0/G1 population to the MFI of the normal reference (Sf9) G0/G1 population. Ploidy of the test sample was then calculated based on the DNA index and the ploidy of the normal reference.

2.8. Karyotype Analysis

Chromosome preparations were obtained from established cultures of Tni-FNL. Vinblastine (5 mg/mL; Sigma, St. Louis, MO, USA) was added to the cells for 2 h prior to harvest and incubated at 27 °C. Cells were treated with hypotonic solution (KCL 0.075M) for 20 min at 37 °C and fixed with methanol: acetic acid 3:1. Slides were prepared at 60% humidity and aged overnight. Pairing was completed using slides that were stained with a trypsin-Giemsa staining technique (GTG). Analyses were performed under an Axio Imager Z2 (Zeiss, Oberkochen, Germany) microscope coupled with a VDS CCD-1300 camera (Genasis, ASI, Carlsbad, CA, USA); images were captured with Spectral Acquisition Band View 7.2 karyotyping software, (Applied Spectral Imaging Inc., Carlsbad, CA, USA).

2.9. NGS De Novo Assembly Methods

The genomic libraries were sequenced on two different sequencing platforms including the PacBio Sequel and RSII systems (Pacific Biosciences, Menlo Park, CA, USA) and the Illumina NextSeq 500 (Illumina, San Diego, CA, USA). We performed de novo assembly of the PacBio sequencing reads using the HGAP4 assembler [14] from SMRT Link software version 4.0.0 (Pacific Biosciences, Menlo Park, CA, USA). The HGAP4 assembly consensus was polished using the Quiver software in the SMRT Link software package. In addition, the Canu v1.4 assembler [15] was used to generate a second set of primary assembly. The Canu assembler was run with all three options of trimming, error correction and assembly. 10X Genomics Supernova v1.2.0 (10X Genomics, Pleasanton, CA, USA) was run iteratively for subsampling in order to find the best genome coverage and optimal assembly results. We subsampled 42× linked reads sequence data and performed de novo assembly.

2.10. Bionano De Novo Assembly

De novo assembly was done using the Bionano Genomics RefAligner version 5122 software (Bionano Genomics, San Diego, CA, USA). First, we merged all the Bionano runs using the merge function of the IrysView. Then the merged molecules set was used for Bionano de novo assembly. The converted Hierarchical Genome Assembly Process (HGAP4) assembly Consensus Map (CMAP) file was also supplied for the error rate estimation. Analysis parameters were given from the optArguments_human.xml. We generated multiple assemblies using the different minimum length cutoffs (150 kb, 180 kb and 210 kb) with two different CMAP-converted fasta assemblies (HGAP4 and Canu). After checking the resulting de novo assemblies, we decided to use the 150 kb minimum length cutoff in conjunction with the HGAP4 fasta file supplied as the CMAP file for our final de novo assembly.

2.11. Bionano Hybrid Assembly

For the step-one hybrid assembly (V1) we used the de novo Bionano assembly with the HGAP4 fasta assembly using the parameters from the aggressive human assembly setting, choosing Nb.BssSI (New England Biolabs, Ipswich, MA, USA) as the enzyme and a threshold p value of 1×10^{-10}. In the two steps of hybrid scaffolding to align Bionano genome maps with PacBio WGS assemblies, the parameters -B2 and -N1 were used to only cut optical mapping assemblies when a conflict was found. For the step-two (V2) version of the hybrid assembly, we merged the mapped CMAP file in the hybrid V1 assembly with the unmapped CMAP file not used in the hybrid assembly with RefAligner version 5122 (Bionano Genomics, San Diego, CA, USA). Then we used the merged CMAP from V1 hybrid with the Canu assembly to carry out the V2 version of the hybrid assembly. The same parameters were used for the V2 hybrid assembly.

2.12. Assembly Error Correction

The final hybrid scaffold assemblies were error-corrected using Pilon [16]. The raw Canu assembly was mapped to Illumina data using the BWA-mem aligner. After mapping, the bam file and Canu raw assembly were supplied to Pilon to perform error correction.

2.13. Transcriptome Assembly

rCorrector [17] was used to remove erroneous k-mers from Illumina paired-end short reads. Adapters and low-quality reads were trimmed using trimmomatic tool. The trimmed pair-end reads were assembled by using trinity assembler (–SS_lib_type FR and –min_kmer_cov 1). The assembly statistics was calculated using Quast. The completeness of the assembly was assessed using BUSCO against Endopterygota database.

2.14. Gene Predictions and Repeat Annotations

The complete genes were predicted from the repeat masked genome using Maker v 2.31.8 pipeline [18] as described in the GC Specific Maker pipeline. After the first initial run of Maker with est2genome [19], the resulting annotation was divided based on GC content as high and low GC data sets. High and low GC datasets along with the original first maker annotations were used to train the SNAP [20] and Augustus [21] HMMs for the gene prediction. In the final run, the assembly was trained against six models, including three from SNAP and three from Augusts, using Maker. The high quality gene models were filtered by choosing Annotation Edit Distance (AED) cut off 0.5 according to the published Maker protocol [22].

2.15. Phylogeny Analysis for Ten Insect Genomes

Genomes of *Bombyx mori* (GCA_000151625.), *Cimex lectularius* (GCA_001460545.1), *Bombus terrestris* (GCF_000214255.1), *Bombus impatiens* (GCF_000188095.1), *Helicoverpa zea* (GCA_002150865.1), *Mamestra configurata* (GCA_002192655.1), *Helicoverpa armigera* (GCA_002156985.1) and *Cimex lectularius* (GCA_000648675.1). were downloaded from NCBI. For *Drosophila*, the BDGP6 version of the genome was used. Nine genomes described above and the *Trichoplusia ni* Tni-FNL assembly were used with Busco [23] to annotate the completeness of single-copy orthologs. We used a total of 250 strict one-to-one orthologs from the 10 species to run the phylogeny analysis. One fasta sequence was generated per species by appending each of the 250 ortholog sequences. The final file containing a single sequence per species was used for multiple sequence alignment by MUSCLE [24]. RAXML [25] was used to generate the maximum likelihood phylogeny from the concatenated multiple sequence alignment using 1000 bootstrap. The resulted Newick formatted tree was plotted using the iTOL [26].

3. Results

3.1. Genome Sequencing and Assembly

To build the assembly, we used a combination of three technologies, including PacBio single-molecule long-read sequencing, Bionano optical genome mapping, and 10X Genomics long linked-reads. For PacBio sequencing, we constructed and sequenced a 20 kb SMRTbell library using 16 SMRT cells on the PacBio RS II. 27.3 Gb of data were generated with a 7.2 kb mean subread length. An additional 20 kb library was constructed using the Swift Biosciences Accel-NGS® XL Library Kit (Swift Biosciences™, Ann Arbor, MI, USA) and was sequenced using 2 SMRT cells on the PacBio RS II, generating 3.1 Gb of data with a 9.7 kb mean subread length. In addition, a 20 kb library was prepared by PacBio and sequenced using 2 SMRT cells on the PacBio Sequel platform, generating an additional 11 Gb of data with a 11 kb mean subread length. In total, 4,236,403 subreads composing 41 Gb were produced by the PacBio platforms, representing approximately $110\times$ coverage of the genome (Table S1). We first performed a contamination check of the PacBio long reads using DeconSeq [27] and found no contaminants in the sequencing reads. We then performed de novo assembly of the PacBio long reads using both the HGAP4 assembler [14] and the Canu v1.4 assembler [15]. The HGAP4 assembly was error corrected and polished using Quiver in the SMRT Link software package. The resulting HGAP4 assembly had 1428 contigs with a total length of 366.3 Mb. The HGAP4 contig N50 size was 939.8 kb and the maximum contig size was 4.35 Mb. The Canu assembly was performed using the options of trimming and error correction and yielded 2101 contigs, with a total size of 408.4 Mb. The Canu N50 contig size was 737.2 kb and the maximum contig size was 6.1 Mb. Since the HGAP4 and Canu v1.4 assemblers utilize different algorithms, the results may differ and be complementary, potentially providing better coverage of the whole genome when both are considered during downstream analysis.

The 10X Genomics linked-reads library was constructed and then sequenced on an Illumina NextSeq 500, producing 918 million 2 × 150 pair-end reads, with an estimated $338\times$ depth of coverage. We ran the 10X Genomics Supernova version 1.2 (10X Genomics, Pleasanton, CA, USA) de novo assembly software [28] and sub-sampled the barcoded reads to get an effective coverage of approximately $42\times$ of the whole genome for an optimal assembly result. The Supernova total scaffold size was 375.8 Mb with a maximum scaffold of 7.83 Mb and N50 size of 1.63 Mb. The Supernova total contig size of 316.7 Mb was comprised of 18,923 contigs with a N50 contig size of 54 kb including 6467 contigs exceeding 10 kb in length. The Supernova assembler produced a much more fragmented assembly, with a total contig size much shorter than that of the PacBio WGS assemblies produced by the HGAP and Canu v1.4 assemblers. Table 1 shows the comparative assembly metrics generated by using QUAST [29] for the three NGS assemblies.

Table 1. Comparison of Tni-FNL de novo assembly statistics.

Types	HGAP Contigs	Canu Contigs	Supernova Contigs
Total contigs	1428	2101	18,923
Contigs ($\geq$1000 bp)	1418	2101	18,196
Contigs ($\geq$10,000 bp)	1323	2097	6467
Contigs ($\geq$25,000 bp)	1101	1780	3592
Contigs ($\geq$50,000 bp)	706	1041	1828
Largest contig (bp)	4,352,893	6,104,320	445,812
Total length (bp)	366,261,337	408,408,011	316,721,011
GC (%)	35.66	35.74	35.33
N50 (bp)	939,843	737,233	54,240
N75 (bp)	421,565	250,244	22,505
L50 (bp)	115	158	1653
L75 (bp)	259	399	3894

3.2. Improved Genome Assembly Using Bionano Optical Mapping Data

To improve the PacBio assemblies and generate hybrid scaffolds representing chromosomal structure, we used Bionano's Irys System to generate optical mapping data. Bionano maps can order and orient sequence fragments to build scaffolds, identify potential chimeric joins in the sequence assembly, and resolve conflicts between a WGS assembly and genome maps [30]. The high-molecular-weight (HMW) DNA molecules were nicked using Nb.BssSI enzyme (New England Biolabs, Ipswich, MA, USA), based on an optimal label density of 14.1 labels per 100 kb, as predicted by the Knickers software (Bionano Genomics, San Diego, CA, USA) from Bionano. We ran this library on the Irys system on 6 flowcells, and merged all 6 runs of molecule data for de novo assembly. The resulting de novo assembly of maps had an average depth of molecule coverage of $61.8\times$ and a total size of 645 Mb in 1272 Bionano maps. Among these 1272 maps, 730 were between 10–500 kb, 469 were between 500–1000 kb, 71 were above 1000 kb and 2 were longer than 2000 kb. The N50 map size was 608 kb. We chose maps greater than 150 kb and combined them with the PacBio WGS assemblies to produce ultra-long hybrid scaffolds using a two-step hybrid approach (Figure 1). We used the Bionano Hybrid Scaffold pipeline (Bionano Genomics, San Diego, CA, USA) [30] to produce "V1" hybrid scaffolds with the HGAP4 WGS assemblies. A total of 301 V1 hybrid scaffolds were produced, with a total assembly size of 328.2 Mb. The N50 of the resulting assembly was 1.74 Mb and the maximum scaffold size was 10.4 Mb. The Bionano software produces a CMAP file as output, which is a raw data view of a molecule set or assembly reporting the label site positions within a genome map. We then merged the mapped CMAP file from the V1 hybrid assembly with the unmapped CMAP files and used Canu v1.4 assembly to carry out a "V2" hybrid assembly. The V2 scaffold reduced the total number of scaffolds from 301 to 280, further increased the N50 size to 2.33 Mb and produced a longer maximum scaffold size of over 12.8 Mb. The total assembled scaffold size increased to 359.1 Mb. This V2 hybrid assembly significantly improved the scaffold N50 size by two-fold compared to the HGAP4 WGS assembly. Additionally, the longest scaffold size in the V2 hybrid assembly is twice the length of the longest contig size in the HGAP4 WGS assembly (Table 2). The gaps (fraction of Ns) in the V2 scaffolds were only 5.5% of the total 359.1 Mb scaffold size.

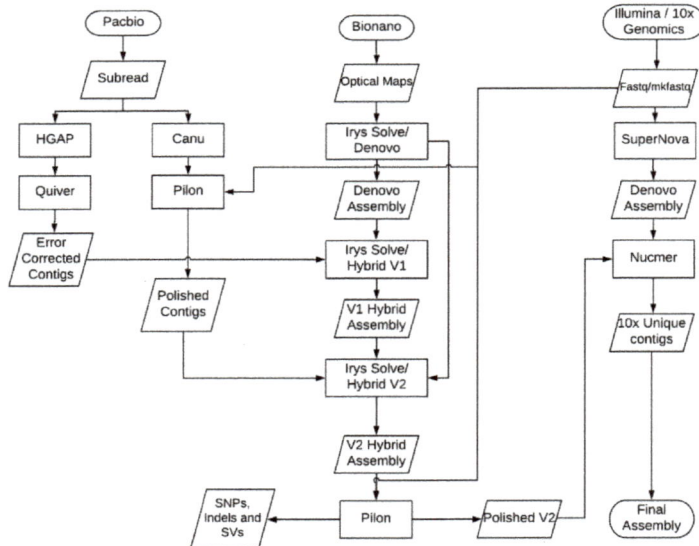

Figure 1. Genome assembly and optical maps hybrid scaffold workflow. The workflow steps for de novo assembly, hybrid scaffold, genome assembly error correction and polishing.

Table 2. Comparison of Tni-FNL genome hybrid scaffold statistics.

Types	Supernova Scaffolds	Hybrid Scaffolds (V1)	Polished Hybrid Scaffolds (V2)
Total scaffolds	12,875	301	280
Scaffolds ($\geq$1000 bp)	12,875	301	280
Scaffolds ($\geq$50,000 bp)	355	301	280
Largest scaffold (bp)	7,830,761	10,389,188	12,760,714
Total scaffold length (bp)	375,813,451	328,208,105	359,075,955
GC (%)	35.33	35.5	35.56
N50 (bp)	1,628,260	1,737,254	2,326,860
N75 (bp)	449,524	1,019,983	1,198,934
L50 (bp)	66	57	44
L75 (bp)	172	120	98
# N's per 100 kb	15,724	2919	5453

3.3. Assembly Conflict Resolution

We identified 789 inconsistent regions when comparing the PacBio WGS assemblies to the Bionano genome maps. We specifically looked for chimeric joins, which are formed when PacBio reads are too short to span across extremely long DNA repeats. These errors would appear as conflicting junctions in the alignment between the PacBio WGS assemblies and Bionano genome maps [30]. The Bionano de novo assembly software reported the conflict regions as alternative consensus maps representing different haplotypes. There was a total of 1029 WGS contigs, of which 637 (61.9% of the total) anchored within the hybrid scaffolds when using a p value of 1×10^{-10} as the cutoff threshold. Among the 789 identified conflicts, 199 chimeric junctions were identified and automatically resolved by the Irys hybrid scaffold software (Figure S1). The high number of remaining unresolved conflicts between the WGS assemblies and Bionano optical maps, as well as a high abundance of short fragment sizes composing the optical maps, suggests the presence of shorter molecules in the Bionano library, and also indicates that the Tni-FNL cell line genome may be highly polymorphic. It was previously reported that insect cell lines used to produce recombinant proteins are cytologically unstable, resulting in varying numbers of chromosomes, depending on the culture history and supplier [31].

3.4. Error Correction of Genome Assembly

The final assembled genome sequences were error-corrected by using two software tools: SMRT analysis resequencing module and Pilon pipeline software [16]. We first mapped the PacBio quality-filtered reads to the hybrid assembly sequences to identify consensus and variant sequences using the PacBio Quiver software. This produced both BAM files and lists of variants in VCF format.

Pilon was used to improve the final hybrid assembly by using read alignments from the Illumina 10X Genomics linked-read data set. Pilon found and fixed 1566 SNPs, 5996 small insertions (consisting of 8848 bases), 11 small deletions (consisting of 3237 bases), 1164 local misassembles, and 2 gaps. This step reduced the total gap size by 26.8 kb in the final hybrid scaffold, and produced a polished final draft genome assembly (Table 2). Of the final 280 scaffolds, 171 scaffolds have sizes greater than 500 kb, 4 scaffolds have sizes less than 100 kb, and the rest of the scaffolds have sizes between 100 kb and 500 kb. (Figure S2).

3.5. Genome Size Estimation Based on K-Mers

To estimate the genome size independently of assembly, we characterized the genome sequence using k-mer histograms, which was computed from the error-corrected reads using the program Jellyfish [32], with word sizes k from 19 to 31. Figure S3 shows the k-mer plot of 1N genome. We also used GenomeScope [33] to profile the genome from the complete set of Illumina short reads. This method gave an estimated haploid genome length of about 328 Mb, and estimated 86.2% of the genome was unique and the overall rate of heterozygosity of the genome was 0.35%, based on k-mer

27 profiling. The lower heterozygosity of the Tni-FNL cell line genome suggests that the Tni-FNL cells were relatively homogenous. This observation is consistent with the recently published Hi5 cells genome sequencing paper, which concluded that the Hi5 cells originated from a single founder cell or a population of homogenous cells, which are different from animal genome.

3.6. Genome Assembly Quality and Completeness Assessment

We first mapped the Illumina 10X Genomics pair-end reads to our Tni-FNL draft genome sequence. Of the total 918.4 million reads, 93.6% was mapped to the 359 Mb draft genome sequence. Only 0.05% of the draft genome had no sequencing coverage from Illumina pair-end reads. This indicates that our Tni-FNL genome assembly is nearly complete.

In addition, we mapped the 10X Genomics Supernova contigs to our Tni-FNL draft genome assembly sequences using Nucmer [34] and produced a mapped BAM file. Of the total 276.35 Mb 10X Genomics Supernova contigs that were longer than 10 kb, only 7.12 Mb were either totally unmapped or partially unmapped, composing only 2.9% of the 10X Genomics Supernova contigs that were not represented in the draft Tni-FNL genome assembly. From the set of 10X Genomics Supernova contigs that were aligned with the draft genome assembly, 99.4% were also aligned concordantly (Figure S4), indicating that the assembly was correct at the local level. Comparing the 10X Genomics Supernova de novo assembly with the Bionano consensus genome map and WGS hybrid assembly, the latter had a much longer N50 scaffold size and better contiguity.

We also compared the transcriptome data of the High Five cell line published in 2016 [7] to our Tni-FNL draft genome assembly. Among the 25,234 assembled transcripts in the High Five cell line, 95.1% were mapped to our Tni-FNL draft genome assembly and 91% were uniquely mapped. This further indicates that our draft genome assembly is correct at the local level and nearly complete.

Compared with the two other lepidopteran genome assemblies, *S. Frugiperda* (358 Mb) [5] and *B. mori* (432 Mb) [4], our genome assembly of the *Trichoplusia ni*-derived Tni-FNL cell line produced much longer N50 contig sizes and far fewer gaps, representing the most contiguous genome assembly for any lepidopteran genome to date (Table 3).

Table 3. Comparison of Tni-FNL genome assembly with other lepidopteran genome assemblies.

Types	Tni-FNL	*Bombyx mori*	*S. frugiperda*
Total length (bp)	359,075,955	481,803,763	358,050,723
Total scaffolds	280	43,462	37,243
Ungapped length (bp)	339,494,557	431,707,935	332,569,779
Scaffold N50 (bp)	2,326,860	4,008,358	53,779
Total contigs	2,043	88,672	>49,244
Contig N50 (bp)	893,993	15,508	7,851
Largest Scaffold (bp)	12,760,714	14,496,184	641,448
Largest Contig (bp)	6,104,547	139,031	234,570
GC (%)	35.56	37.70	32.97
Gap size (bp)	19,581,398	50,095,828	25,480,944

3.7. Determination of Cell Line Ploidy and Karyotype Analysis

Previous analysis of the chromosome content of lepidopteran cell lines showed a lack of consistency in the ploidy of these cells. Sf9 cell lines were shown to have a mixed population of diploid and tetraploid cells that varied in their ratios even when cells were cloned out [35]. *Bombyx mori* cell lines were mostly diploid but by examining a number of lepidopteran lines, ploidy was found to be highly variable [36]. While no characterization of *T. ni* cell line ploidy has been published so far, based on cell size data, we speculated that it was very probable that these cell lines were tetraploid. Propidium iodide staining confirmed most Tni-FNL cells contain a 4n DNA content (Figure S5), similar to the profiles previously observed for selected tetraploid Sf9 cells. In contrast, our Sf9 control cell line (derived from ATCC CRL-1711) is mostly diploid.

Lepidopteran chromosomes have been shown to be holokinetic, lacking centromeric structures, which makes them highly prone to chromosome fragmentation and complex karyotyping [37,38]. Due to this lack of centromeres, different karyotypic staining techniques were tried. In the end, pairing was completed using slides that were stained with a trypsin-Giemsa staining technique (GTG). The GTG banding provided slight differences in the structure of the chromosomes that were used to match similar chromosomes and fragments. As shown in Figure S6, karyotyping of the Tni-FNL cells was consistent with the suggestion of a tetraploid genome with visible chromosome fragmentation. The average chromosome number over several spreads was 130. In the context of the putative tetraploid state of Tni-FNL, these data are consistent with the previous identification of 28–30 chromosomes for most lepidopteran organisms. The chromosome number appears to be lower than previous studies on Sf9 cells, which suggests much higher levels of chromosome fragmentation in those cells [35].

3.8. Analysis of GC Content and CpG Islands

CpG islands, which are clusters of CpG dinucleotides in GC-rich regions, represent important features in insect genomes. Analysis of GC content and CpG islands can help identify the role of CpG islands in gene regulation and evolution [39]. A previous study found uniformity in GC content and CpG islands among the lepidopteran insects [10]. Our draft assembly of the Tni-FNL genome shows a GC composition of 35.7% of the total bases. This composition is very similar to the closely related species *S. Frugiperda* (33.0%), and *B. mori* (37.7%) (Figure S7) and further supports the notion that closely related organisms share common features at the genomic level.

To study CpG island distribution in the genome, we used EMBOSS [40], which enabled identification of 4426 CpG islands with a total length of approximately 2.8 Mb (0.8% of the genome) from a total of 280 final scaffold assemblies. Greater than 83.4% of the CpG islands were identified when using a genomic window length of 200–800 bases. By comparing the GC content distribution against the occurrence of CpG islands, we found that the CpG islands occur in the genome where the GC content ranges between 51% and 75%. The average GC content within the CpG islands is 63.6%, while the average GC content of the whole genome assembly is 35.7%. This is similar to that of the other lepidopteran insects, such as *B. mori* and *S. Frugiperda* (Figure S8). The uniformity in GC percentage among the lepidopteran insects confirms the previous observation and supports the notion that closely related insect species share similar genomic patterns [4,41].

3.9. Analysis of Repeat Elements Including Endogenous Viral Elements

To identify repetitive elements within the assembled genome, we ran Repeat Modeler (http://repeatmasker.org) software for de novo modeling of the Tni-FNL genome. A model library was constructed by using the Repbase [42] known repeat library. Repeat masking was accomplished using Repeat Masker with the model library generated by the Repeat Modeler. All identified repeats were annotated using Repbase classification. Our analysis revealed a total of 517,939 standard genome repetitive elements, of which 127,344 were simple repeats, 366,173 were interspersed repeats, 42,457 were retro elements including LINEs, SINEs, and LTRS, and approximately 2,986 were DNA elements. In total, 18.8% of the Tni-FNL assembly (67.5 Mb) was found to be repetitive and a major fraction of the repeats were classified as interspersed repetitive elements (Table S2).

The percentage of the Tni-FNL genome made up of repetitive elements (Figure S7) was consistent with the numbers reported for other insect genomes, including other lepidoptera such as *S. Frugiperda* (20.28%) [5] and *B. mori* (43.6%) [4], as well as other related insects such as *C. lectularius* (31.65%) [43], *B. terrestris* (14.8%) [44], *B. impatiens* (17.9%) [44], and *D. melanogaster* (20%). We also searched the Tni-FNL genome sequence for endogenous viral elements (EVEs) including endogenous retroviruses (ERVs) among mammalian genomes, and found that 3183 bases of unique EVEs elements hit the genome. Transcriptionally active EVEs have been suggested to confer protection or tolerance against related exogenous viruses [45,46].

3.10. Gene Prediction and Functional Annotation

We predicted 14,101 gene models in the Tni-FNL genome based on the Maker2 [18] pipeline, which utilizes known proteins, expressed-sequence tags (ESTs), or assembled transcripts to predict gene models. The High Five cell line (BTI-Tn-5B1-4) transcriptome sequencing data, containing 25,234 transcripts [7], and the *B. mori* insect annotation, containing 19,559 protein sequences [4], were downloaded from GenBank and used as input training files for the Maker pipeline for gene annotation. After the first initial run with est2genome [19], the resulting annotation data sets were provided to the Snap [20] and Augustus [21] pipelines for gene structure prediction. The initial total number of predicted gene models in Tni-FNL genome was 41,078. We filtered false positive gene models by choosing Annotation Edit Distance (AED) less than 0.5. AED is a distance measure that summarizes the congruency of each annotation with its supporting evidence according to the published Maker protocol [22]. This produced 14,101 final gene models. The total count of final gene models of Tni-FNL is similar to the total genes predicted in closely related species such as *S. Frugiperda* (11,595) and *B. mori* (16,424), *D. melanogaster* (17,746) as well as to Hi5 germ line cell (14,037).

To perform functional analysis of the predicted genes, we ran an InterPro [47] search against the InterPro consortium databases including Pfam, PROSITE, TIGRFAMs, CDD and 10 other databases based on homology searches. The InterPro search resulted in 13,143 protein coding genes (93.2% of total). Protein sequences were classified into families and assigned domains or functional sites. Among the 13,143 protein sequences, 8680 (66%) protein sequences were assigned Gene ontology (GO) terms, and 10,846 (82.5%) had a Pfam domain assignment. The GO terms were summarized into three main GO categories-biological processes, cellular components and molecular functions. We found a total of 884 biological processes, 316 cellular components and 952 molecular functions for the predicted gene models in the Tni-FNL genome (Table S3). The most abundant (top 10) subcategory genes are selected and shown in Figure 2.

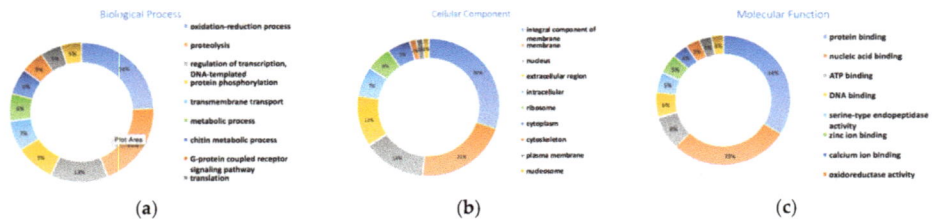

Figure 2. Gene ontology classification of the genes predicted. Gene Ontology (GO) classification of the predicted genes. Only the most abundance ones are displayed. (**a**) biological processes, (**b**) cellular components, (**c**) molecule functions.

In order to assess the quality of the genome assembly, we used a protocol developed by the Maker authors, who state that if 90% of the annotations have an annotated estimated distance (AED) less than 0.5 and more than 50% of the proteome contains a recognizable protein domain, then the genome can be defined as well annotated [22]. Our genome assembly predicts for 13,143 protein sequences that have recognizable domains or protein families assigned, which makes up 93.2% of the total proteome. The full set of 14,101 gene models all have an annotated estimated distance (AED) less than 0.5. These data strongly indicate that this genome is well annotated. It is worth noting that by incorporating RNA-seq data, as well as the annotated silkworm *B. mori* assembly [4] in our training data set, we have created an expansive gene model set, which we believe produced a more complete set of gene annotations for the *Trichoplusia ni* genome.

In a comparison of GO category genes of the *T. ni* genome with *B. mori* and *D. melanogaster*, the majority of these are consistent among the three species (Figure 3, Table S4). This result is consistent with earlier findings that insects share a common set of genes to maintain their integrity though their evolutionary pattern, while there is a selection for a set of genes or protein families among them conferring uniqueness to each insect [4].

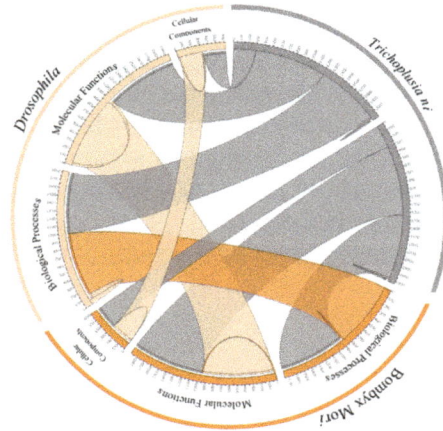

Figure 3. Functional annotation of the Tni-FNL (*Trichoplusia ni*), *B. mori* and *D. melanogaster* results comparison. The circos plot describes the shared cellular components, molecular functions and biological processes among the three species.

We also compared the predicted gene structures of Tni-FNL with closely related species such as *S. frugiperda* (11,595 predicted genes) and *B. mori* (16,424 predicted genes). The total number of exons in the predicted genes of Tni-FNL (105,550) were between that of *S. frugiperda* (64,725) and *B. mori* (197,632). The number of exons per transcript is very similar among the three species (5.6, 7 and 8 respectively among *T.ni*, *S.f.* and *B. mori*) (Table S5). Approximately one-third of the Tni-FNL genome is comprised of genic regions and 6% by coding sequences. This is consistent with the well annotated *B. mori* annotation published in 2017 by NCBI (Annotation release 102), which shows that approximately 57% of the genome is genic and 8% contains coding sequences.

3.11. Orthologs and Phylogenetic Analysis

For the ortholog analysis, we searched the OrthoDB6 [48] database using RefSeq genes from 10 genomes, including Tni-FNL (*Trichoplusia ni*), *Spodoptera frugiperda*, *Bombyx mori*, *Cimex lectularius*, *Bombus terrestris*, *Bombus impatiens*, *Helicoverpa zea*, *Helicoverpa armigera*, *Mamestra configurata*, and *Drosophila melanogaster*. BUSCO [23], a known benchmarking approach for assessing single-copy orthologs conserved among species, was used to annotate the completeness of single-copy orthologs in the above genomes. We found 2175 complete and single-copy orthologs in *Trichoplusia ni* (89.1%) out of 2442 total orthologs of the *Holometabola* lineage. This provides further evidence of high accuracy and completeness of our Tni-FNL cell line genome assembly (Figure 4).

In addition, we also found that the Tni-FNL contained a much higher number of complete and duplicated orthologs (3.5% of total 2442 orthologs of the *Holometabola* lineage) than the other 9 species in the comparison. This indicates the Tni-FNL cell line has higher levels of chromosome duplications, which suggests that Tni-FNL cells may be mostly tetraploid.

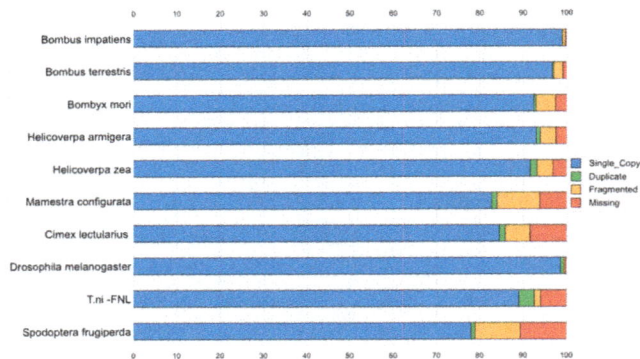

Figure 4. BUSCO assessment results of orthologs among 10 species. Colors refer to the percentage of the complete single-copy orthologs (blue), complete duplicated orthologs (green), fragmented or incomplete orthologs (orange), and missing orthologs (red).

Our finding further supports previous studies showing that majority of genes have orthologous relationships across species in the same lineage, and the lineage-specific orthologs are likely to play important roles in lineage-specific biological traits [44]. In addition, we used the set of 250 strict one-to-one orthologs from all of the above 10 species for phylogenetic analysis. RAXML [25] was used to generate the maximum likelihood phylogeny from the concatenated multiple-sequence alignments. The gene content matrices were analyzed using the BINGAMMA model in RAXML. The analysis results were entirely in agreement with the accepted topology of insect relationships among the selected 10 species [26,47] (Figure 5).

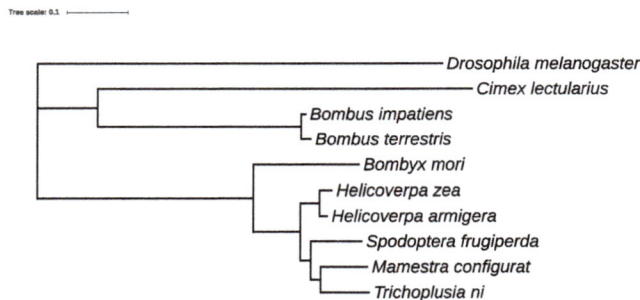

Figure 5. Phylogenetic analysis of *Trichoplusia ni* and closely related insect genomes. Phylogenetic analyses from 10 species including 6 lepidopterans. It depicts the relationship of *Trichoplusia ni* with the other nine insects. Maximum Likelihood tree based on a genome-wide one-to-one orthologs from 10 species. The scale bar denotes substitutions per site.

4. Discussion

By combining PacBio single-molecule long-read sequencing with Bionano optical genome mapping and 10X Genomics long linked-reads technologies, we were able to produce a high-quality genome assembly of the Tni-FNL cell line genome. Since lepidopteran chromosomes are prone to chromosome fragmentation and complex karyotyping [37,38], assembly of a lepidopteran host genome presented a major challenge. With an average PacBio sequencing read length greater than 10 kb, the reads could easily span most repetitive elements and were unambiguously placed on the correct chromosomes, which enabled us to build a highly contiguous assembly. Two sets of WGS assemblies from the HGAP4 and Canu assemblers were generated in this process. We further improved the

WGS assemblies with integration of Bionano genome maps to build hybrid scaffolds. Genome maps helped identify chimeric contigs and fixed mis-assemblies and redundancies present in the WGS contigs. We improved the hybrid scaffold results by using the long linked-reads generated from the 10X Genomics Chromium platform. The purpose of using barcoded long linked-reads that originated from larger, single molecules of DNA was to replace the approach of using costly BAC clone or pooled fosmid clone libraries. The 10X linked-reads assemblies can effectively measure the contiguity and completeness of the hybrid scaffolds produced from the PacBio and Bionano data and help to identify connection errors found in the PacBio-Bionano hybrid assembly. By combining single-molecule sequencing with complementary technologies such as optical genome mapping and 10X linked-reads, we produced a high-quality genome assembly, which comprises 359 million base pairs with fewer than 5.5% of gaps. It represents one of the most contiguous draft assemblies of a lepidopteran host genome to date.

To further assess the genome assembly quality and elucidate gene functions in this lepidopteran host, we performed the gene prediction and comparative analysis with other insect genomes, as well as utilized the transcriptome sequencing data for this cell line and data generated from previous studies [6,7]. Comparative analyses of orthologs from the Tni-FNL (*Trichoplusia ni*) genome and other lepidopteran hosts confirmed the previous study findings that insect genomes share a common set of genes to maintain their integrity through their evolution [44]. The total number of repeat elements identified in the Tni-FNL genome is very similar to those reported in other lepidopteran insect genomes. The results offer insights for further studies on how changes in the degree of repeat regions are involved in maintaining genome integrity among insert genomes.

The predicted genes of Tni-FNL identified in this study also provide additional resources for studying genetic variations and genome evolution in insect genomes. As previous studies suggested, the full-genome sequences from multiple species can complement each other by clarifying gene function and organization. In addition, this work will enable efforts to develop system biology tools to improve the utility of the Tni-FNL cell line for protein production. Recent reports have demonstrated for the first time the ability to genetically engineer *Trichoplusia* cell lines using the CRISPR/Cas9 system [49]. High-quality genome modification requires detailed genomic information to ensure high efficiency of targeted modifications and reduction in unwanted off-target effects. Using the high-quality genome sequence and newly developed engineering tools, we believe it will be possible to begin to make modifications to Tni-FNL, which will ultimately improve the quality and lower the cost of therapeutic protein production using this system.

Supplementary Materials: The following are available online at http://www.mdpi.com/2073-4425/10/2/79/s1, Figure S1: Overview of the comparison of PacBio assemblies to Bionano genome maps, Figure S2: Hybrid assembly scaffolds size distribution, Figure S3: K-mer counts plot, Figure S4: Dot plots display alignment of 10X Genomics Supernova contigs to the final hybrid scaffolds, Figure S5: Flow cytometric analysis of DNA content of insect cell lines, Figure S6: Image and karyotype of Tni-FNL cell line, Figure S7: GC content and repeat elements comparison among three species, Figure S8: Identification of CpG islands in the Tni-FNL genome; Table S1: Data Generated from Three Different Technologies, Table S2: Repeat elements identified in the Tni-FNL genome sequence, Table S3: Gene ontology classification of the genes predicted from the Tni-FNL genome assembly, Table S4: Comparison of shared GO category genes from Tni-FNL, *B. mori* and *D. melanogaster*, Table S5: Transcript Structure Comparison between Tni-FNL, *S. frugiperda* and *B. mori*.

Author Contributions: D.E. and Y.Z. initiated and planned the project. Y.Z., B.T., C.R. and M.M. designed the overall sequencing experiments. C.G. carried out DNA production, M.D. carried out insect cell expression, V.R. isolated and adapted the Tni-FNL strain, K.T.C. carried out flow cytometry analysis, S.B. performed karyotyping analysis, C.R., M.M., Y.K., S.G., and P.K. prepared NGS libraries and Bionano Optical mapping experiments. J.S., O.G., T.S., and C.H. performed library QC and sequencing, S.O. provided Bionano experimental design and helped optical map data quality assessment, K.T. performed genome assembly and annotation. Y.Z. supervised data analysis, performed genome assembly, annotation, comparative analyses with other insect genomes, X.C. helped assembly software tools, R.S. and D.S. provided sequencing and informatics support for this project. B.T. oversaw sequencing experiments. D.E. designed insect cell experiments, analyzed karyotype data, and edited the manuscript. Y.Z. wrote the manuscript, assisted by D.E., C.R., M.M., and K.T. All authors reviewed and approved the final manuscript.

Funding: This work was funded in whole or in part with Federal funds from the National Cancer Institute, National Institutes of Health, under Contract No. HHSN261200800001E. The content of this publication does not necessarily reflect the views or policies of the Department of Health and Human Services, nor does mention of trade names, commercial products, or organizations imply endorsement by the U.S. Government. This research was supported (in part) by the National Institutes of Health.

Acknowledgments: We would like to thank the CCR Sequencing Facility at the Frederick National Laboratory for Cancer Research. We are grateful for the support received from Dwight Nissley and Jack Collins from the Frederick National Laboratory for Cancer Research. Thanks to Kristina Weber and Roberto Lleras from PacBio for providing PacBio software help, and many thanks to Weining Xu, Yan Guo, Christine Lambert and Primo Baybayan at PacBio for producing the Sequel data. Special thanks go to Michael Schatz from Johns Hopkins University, who provided valuable input and comment regarding K-mer based genome analysis. We also wish to thank Xiao-Dong Su and Kai Yu at Peking University for providing us access to their Hi5 transcriptome data.

Conflicts of Interest: The authors declare that they have no competing interests.

References

1. Davis, T.R.; Wickham, T.J.; McKenna, K.A.; Granados, R.R.; Shuler, M.L.; Wood, H.A. Comparative recombinant protein production of eight insect cell lines. *In Vitro Cell Dev. Biol. Anim.* **1993**, *29A*, 388–390. [CrossRef] [PubMed]

2. Cox, M.M. Recombinant protein vaccines produced in insect cells. *Vaccine* **2012**, *30*, 1759–1766. [CrossRef] [PubMed]

3. Felberbaum, R.S. The baculovirus expression vector system: A commercial manufacturing platform for viral vaccines and gene therapy vectors. *Biotechnol. J.* **2015**, *10*, 702–714. [CrossRef] [PubMed]

4. Mita, K. The genome sequence of silkworm, bombyx mori. *DNA Res.* **2004**, *11*, 27–35. [CrossRef] [PubMed]

5. Kakumani, P.K.; Malhotra, P.; Mukherjee, S.K.; Bhatnagar, R.K. A draft genome assembly of the army worm, spodoptera frugiperda. *Genomics* **2014**, *104*, 134–143. [CrossRef] [PubMed]

6. Chen, Y.R.; Zhong, S.; Fei, Z.; Gao, S.; Zhang, S.; Li, Z.; Wang, P.; Blissard, G.W. Transcriptome responses of the host trichoplusia ni to infection by the baculovirus autographa californica multiple nucleopolyhedrovirus. *J. Virol.* **2014**, *88*, 13781–13797. [CrossRef] [PubMed]

7. Yu, K.; Yu, Y.; Tang, X.; Chen, H.; Xiao, J.; Su, X.D. Transcriptome analyses of insect cells to facilitate baculovirus-insect expression. *Protein Cell* **2016**, *7*, 373–382. [CrossRef] [PubMed]

8. Wickham, T.J.; Davis, T.; Granados, R.R.; Shuler, M.L.; Wood, H.A. Screening of insect cell lines for the production of recombinant proteins and infectious virus in the baculovirus expression system. *Biotechnol. Prog.* **1992**, *8*, 391–396. [CrossRef] [PubMed]

9. Eid, J.; Fehr, A.; Gray, J.; Luong, K.; Lyle, J.; Otto, G.; Peluso, P.; Rank, D.; Baybayan, P.; Bettman, B.; et al. Real-time DNA sequencing from single polymerase molecules. *Science* **2009**, *323*, 133–138. [CrossRef]

10. Lam, E.T.; Hastie, A.; Lin, C.; Ehrlich, D.; Das, S.K.; Austin, M.D.; Deshpande, P.; Cao, H.; Nagarajan, N.; Xiao, M.; et al. Genome mapping on nanochannel arrays for structural variation analysis and sequence assembly. *Nat. Biotechnol.* **2012**, *30*, 771–776. [CrossRef]

11. Zheng, G.X.; Lau, B.T.; Schnall-Levin, M.; Jarosz, M.; Bell, J.M.; Hindson, C.M.; Kyriazopoulou-Panagiotopoulou, S.; Masquelier, D.A.; Merrill, L.; Terry, J.M.; et al. Haplotyping germline and cancer genomes with high-throughput linked-read sequencing. *Nat. Biotechnol.* **2016**, *34*, 303–311. [CrossRef] [PubMed]

12. Phillippy, A.M. New advances in sequence assembly. *Genome Res.* **2017**, *27*, xi–xiii. [CrossRef] [PubMed]

13. Fu, Y.; Yang, Y.; Zhang, H.; Farley, G.; Wang, J.; Quarles, K.A.; Weng, Z.; Zamore, P.D. The genome of the hi5 germ cell line from trichoplusia ni, an agricultural pest and novel model for small rna biology. *eLife* **2018**, *7*. [CrossRef] [PubMed]

14. Chin, C.-S.; Alexander, D.H.; Marks, P.; Klammer, A.A.; Drake, J.; Heiner, C.; Clum, A.; Copeland, A.; Huddleston, J.; Eichler, E.E.; et al. Nonhybrid, finished microbial genome assemblies from long-read smrt sequencing data. *Nat. Methods* **2013**, *10*, 563–569. [CrossRef] [PubMed]

15. Koren, S.; Walenz, B.P.; Berlin, K.; Miller, J.R.; Bergman, N.H.; Phillippy, A.M. Canu: Scalable and accurate long-read assembly via adaptive k -mer weighting and repeat separation. *Genome Res.* **2017**, *27*, 722–736. [CrossRef] [PubMed]

16. Walker, B.J.; Abeel, T.; Shea, T.; Priest, M.; Abouelliel, A.; Sakthikumar, S.; Cuomo, C.A.; Zeng, Q.; Wortman, J.; Young, S.K.; et al. Pilon: An integrated tool for comprehensive microbial variant detection and genome assembly improvement. *PLoS ONE* **2014**, *9*, e112963. [CrossRef] [PubMed]

17. Song, L.; Florea, L. Rcorrector: Efficient and accurate error correction for illumina rna-seq reads. *GigaScience* **2015**, *4*, 48. [CrossRef]

18. Holt, C.; Yandell, M. Maker2: An annotation pipeline and genome-database management tool for second-generation genome projects. *BMC Bioinform.* **2011**, *12*, 491. [CrossRef]

19. Mott, R. Est_genome: A program to align spliced DNA sequences to unspliced genomic DNA. *Bioinformatics* **1997**, *13*, 477–478. [CrossRef]

20. Korf, I. Gene finding in novel genomes. *BMC Bioinform.* **2004**, *5*, 59. [CrossRef]

21. Stanke, M.; Morgenstern, B. Augustus: A web server for gene prediction in eukaryotes that allows user-defined constraints. *Nucleic Acids Res.* **2005**, *33*, W465–W467. [CrossRef] [PubMed]

22. Campbell, M.S.; Holt, C.; Moore, B.; Yandell, M. Genome annotation and curation using maker and maker-p. *Curr. Protoc. Bioinform.* **2014**, *48*, 1–39.

23. Simão, F.A.; Waterhouse, R.M.; Ioannidis, P.; Kriventseva, E.V.; Zdobnov, E.M. Busco: Assessing genome assembly and annotation completeness with single-copy orthologs. *Bioinformatics* **2015**, *31*, 3210–3212. [CrossRef] [PubMed]

24. Edgar, R.C. Muscle: Multiple sequence alignment with high accuracy and high throughput. *Nucleic Acids Res.* **2004**, *32*, 1792–1797. [CrossRef] [PubMed]

25. Stamatakis, A. Raxml-vi-hpc: Maximum likelihood-based phylogenetic analyses with thousands of taxa and mixed models. *Bioinformatics* **2006**, *22*, 2688–2690. [CrossRef] [PubMed]

26. Letunic, I.; Bork, P. Interactive tree of life (itol) v3: An online tool for the display and annotation of phylogenetic and other trees. *Nucleic Acids Res.* **2016**, *44*, W242–W245. [CrossRef]

27. Schmieder, R.; Edwards, R. Fast identification and removal of sequence contamination from genomic and metagenomic datasets. *PLoS ONE* **2011**, *6*, e17288. [CrossRef]

28. Weisenfeld, N.I.; Kumar, V.; Shah, P.; Church, D.M.; Jaffe, D.B. Direct determination of diploid genome sequences. *Genome Res.* **2017**, *27*, 757–767. [CrossRef]

29. Gurevich, A.; Saveliev, V.; Vyahhi, N.; Tesler, G. Quast: Quality assessment tool for genome assemblies. *Bioinformatics* **2013**, *29*, 1072–1075. [CrossRef]

30. Shelton, J.M.; Coleman, M.C.; Herndon, N.; Lu, N.; Lam, E.T.; Anantharaman, T.; Sheth, P.; Brown, S.J. Tools and pipelines for bionano data: Molecule assembly pipeline and fasta super scaffolding tool. *BMC Genom.* **2015**, *16*, 734. [CrossRef]

31. Schmutz, J.; Wheeler, J.; Grimwood, J.; Dickson, M.; Yang, J.; Caoile, C.; Bajorek, E.; Black, S.; Chan, Y.M.; Denys, M.; et al. Quality assessment of the human genome sequence. *Nature* **2004**, *429*, 365–368. [CrossRef] [PubMed]

32. Marcais, G.; Kingsford, C. A fast, lock-free approach for efficient parallel counting of occurrences of k-mers. *Bioinformatics* **2011**, *27*, 764–770. [CrossRef] [PubMed]

33. Vurture, G.W.; Sedlazeck, F.J.; Nattestad, M.; Underwood, C.J.; Fang, H.; Gurtowski, J.; Schatz, M.C. Genomescope: Fast reference-free genome profiling from short reads. *Bioinformatics* **2017**, *33*, 2202–2204. [CrossRef] [PubMed]

34. Kurtz, S.; Phillippy, A.; Delcher, A.L.; Smoot, M.; Shumway, M.; Antonescu, C.; Salzberg, S.L. Versatile and open software for comparing large genomes. *Genome Biol.* **2004**, *5*, R12. [CrossRef] [PubMed]

35. Jarman-Smith, R.F.; Mannix, C.; Al-Rubeai, M. Characterisation of tetraploid and diploid clones of spodoptera frugiperda cell line. *Cytotechnology* **2004**, *44*, 15–25. [CrossRef] [PubMed]

36. Lery, X.; Charpentier, G.; Belloncik, S. DNA content analysis of insect cell lines by flow cytometry. *Cytotechnology* **1999**, *29*, 103–113. [CrossRef] [PubMed]

37. d'Alencon, E.; Sezutsu, H.; Legeai, F.; Permal, E.; Bernard-Samain, S.; Gimenez, S.; Gagneur, C.; Cousserans, F.; Shimomura, M.; Brun-Barale, A.; et al. Extensive synteny conservation of holocentric chromosomes in lepidoptera despite high rates of local genome rearrangements. *Proc. Natl. Acad. Sci. USA* **2010**, *107*, 7680–7685. [CrossRef] [PubMed]

38. Lynn, D.E. Development and characterization of insect cell lines. *Cytotechnology* **1996**, *20*, 3–11. [CrossRef] [PubMed]

39. Lequime, S.; Lambrechts, L. Discovery of flavivirus-derived endogenous viral elements inanophelesmosquito genomes supports the existence ofanopheles-associated insect-specific flaviviruses. *Virus Evol.* **2017**, *3*, vew035. [CrossRef]

40. Rice, P.; Longden, I.; Bleasby, A. Emboss: The european molecular biology open software suite. *Trends Genet.* **2000**, *16*, 276–277. [CrossRef]

41. Han, L.; Su, B.; Li, W.-H.; Zhao, Z. Cpg island density and its correlations with genomic features in mammalian genomes. *Genome Biol.* **2008**, *9*, R79. [CrossRef] [PubMed]

42. Jurka, J. Repeats in genomic DNA: Mining and meaning. *Curr. Opin. Struct. Biol.* **1998**, *8*, 333–337. [CrossRef]

43. Rosenfeld, J.A.; Reeves, D.; Brugler, M.R.; Narechania, A.; Simon, S.; Durrett, R.; Foox, J.; Shianna, K.; Schatz, M.C.; Gandara, J.; et al. Genome assembly and geospatial phylogenomics of the bed bug cimex lectularius. *Nat. Commun.* **2016**, *7*, 10164. [CrossRef] [PubMed]

44. Sadd, B.M.; Barribeau, S.M.; Bloch, G.; de Graaf, D.C.; Dearden, P.; Elsik, C.G.; Gadau, J.; Grimmelikhuijzen, C.J.; Hasselmann, M.; Lozier, J.D.; et al. The genomes of two key bumblebee species with primitive eusocial organization. *Genome Biol.* **2015**, *16*, 76. [CrossRef] [PubMed]

45. Flegel, T.W. Hypothesis for heritable, anti-viral immunity in crustaceans and insects. *Biol. Direct* **2009**, *4*, 32. [CrossRef] [PubMed]

46. Holmes, E.C. The evolution of endogenous viral elements. *Cell Host Microbe* **2011**, *10*, 368–377. [CrossRef] [PubMed]

47. Finn, R.D.; Attwood, T.K.; Babbitt, P.C.; Bateman, A.; Bork, P.; Bridge, A.J.; Chang, H.-Y.; Dosztányi, Z.; El-Gebali, S.; Fraser, M.; et al. Interpro in 2017—beyond protein family and domain annotations. *Nucleic Acids Res.* **2016**, *45*, D190–D199. [CrossRef]

48. Zdobnov, E.M.; Tegenfeldt, F.; Kuznetsov, D.; Waterhouse, R.M.; Simao, F.A.; Ioannidis, P.; Seppey, M.; Loetscher, A.; Kriventseva, E.V. Orthodb v9.1: Cataloging evolutionary and functional annotations for animal, fungal, plant, archaeal, bacterial and viral orthologs. *Nucleic Acids Res.* **2017**, *45*, D744–D749. [CrossRef]

49. Mabashi-Asazuma, H.; Jarvis, D.L. Crispr-cas9 vectors for genome editing and host engineering in the baculovirus-insect cell system. *Proc. Natl. Acad. Sci. USA* **2017**, *114*, 9068–9073. [CrossRef]

genes

MDPI

Article

De Novo Assembly of Two Swedish Genomes Reveals Missing Segments from the Human GRCh38 Reference and Improves Variant Calling of Population-Scale Sequencing Data

Adam Ameur [1,*], Huiwen Che [1], Marcel Martin [2], Ignas Bunikis [1], Johan Dahlberg [3], Ida Höijer [1], Susana Häggqvist [1], Francesco Vezzi [2], Jessica Nordlund [3], Pall Olason [4], Lars Feuk [1] and Ulf Gyllensten [1]

[1]   Science for Life Laboratory, Department of Immunology, Genetics and Pathology, Uppsala University, 752 36 Uppsala, Sweden; chehuiw@hotmail.com (H.C.); ignas.bunikis@igp.uu.se (I.B.); ida.hoijer@igp.uu.se (I.H.); susana.haggqvist@igp.uu.se (S.H.); lars.feuk@igp.uu.se (L.F.); ulf.gyllensten@igp.uu.se (U.G.)
[2]   Science for Life Laboratory, Department of Biochemistry and Biophysics (DBB), Stockholm University, 114 19 Stockholm, Sweden; marcel.martin@scilifelab.se (M.M.); francesco.vezzi@scilifelab.se (F.V.)
[3]   Science for Life Laboratory, Department of Medical Sciences, Molecular Medicine, Uppsala University, 752 36 Uppsala, Sweden; johan.dahlberg@medsci.uu.se (J.D.); jessica.nordlund@medsci.uu.se (J.N.)
[4]   Science for Life Laboratory, Department of Cell and Molecular Biology, Uppsala University, 752 36 Uppsala, Sweden; pallolason@gmail.com
*   Correspondence: adam.ameur@igp.uu.se

Received: 28 August 2018; Accepted: 5 October 2018; Published: 9 October 2018

Abstract: The current human reference sequence (GRCh38) is a foundation for large-scale sequencing projects. However, recent studies have suggested that GRCh38 may be incomplete and give a suboptimal representation of specific population groups. Here, we performed a de novo assembly of two Swedish genomes that revealed over 10 Mb of sequences absent from the human GRCh38 reference in each individual. Around 6 Mb of these novel sequences (NS) are shared with a Chinese personal genome. The NS are highly repetitive, have an elevated GC-content, and are primarily located in centromeric or telomeric regions. Up to 1 Mb of NS can be assigned to chromosome Y, and large segments are also missing from GRCh38 at chromosomes 14, 17, and 21. Inclusion of NS into the GRCh38 reference radically improves the alignment and variant calling from short-read whole-genome sequencing data at several genomic loci. A re-analysis of a Swedish population-scale sequencing project yields > 75,000 putative novel single nucleotide variants (SNVs) and removes > 10,000 false positive SNV calls per individual, some of which are located in protein coding regions. Our results highlight that the GRCh38 reference is not yet complete and demonstrate that personal genome assemblies from local populations can improve the analysis of short-read whole-genome sequencing data.

Keywords: de novo assembly; SMRT sequencing; GRCh38; human reference genome; human whole-genome sequencing; population sequencing; Swedish population

1. Introduction

Due to advances in DNA sequencing technologies, whole genome sequencing (WGS) has become an established method to study human genetic variation at a population scale. Large human WGS projects have been initiated in several countries and geographic regions [1–6], in some cases comprising 10,000 individuals or more [7,8]. These genome projects will provide a wealth of information for future research on human genetics, evolution, and disease. Today, the vast majority of human WGS is performed using short-read Illumina sequencing technology, and requires an alignment of the sequence reads to a human reference sequence. The gold standard reference is the GRCh38 release from 2013, which is based on DNA from multiple donors and intended to represent a pan-human genome, rather than a single individual or population group [9]. However, the current GRCh38 reference might not be optimal in the context of population specific WGS projects, and more information could be gained from WGS data by instead using local references genomes, tailored to a specific country or population. For instance, the de novo assembly of 150 Danish individuals based on Illumina mate-pair sequencing have strengthened the hypothesis that regional reference genomes can increase the power of association studies and improve precision medicine [10]. Since Illumina's technology is limited by short read lengths and amplification biases [11], it is not a viable alternative for creating human de novo assemblies comparable to GRCh38 in terms of completeness and contiguity.

A number of sequencing technologies have emerged that are capable of reading very long DNA molecules without prior amplification. These methods can resolve complex regions of the human genome, such as GC-rich regions or repeats, which are difficult to determine with amplification-based and short-read approaches [12]. In particular, PacBio's single-molecule real-time (SMRT) sequencing technology has proven to be an excellent method for de novo genome assembly. In 2015, the first human de novo SMRT sequencing project was reported; the assembly of the CHM1 cell line derived from a haploid hydatidiform mole [13]. Since then, a handful of human genomes have been assembled using combinations of long-read, linked-read, and optical mapping technologies [14–17], including the AK1 cell line originating from a Korean individual [15] and the HX1 genome originating from a healthy male Han Chinese [14]. These personal genomes have been assembled to a high level of completeness. For example, the AK1 assembly has a contig N50 size of 17.9 megabases (Mb) and scaffold N50 size of 44.8 Mb, with eight chromosome arms resolved into single scaffolds [15]. The contigs assembled from SMRT sequence data, as opposed to most assemblies based on Illumina data, are completely gap-free and contain no ambiguous bases (represented by N's). Approximately 20,000 structural variations (SVs) are detected by SMRT sequencing of a human individual [14,15] and a majority of these SVs are missed by analyses of short-read Illumina WGS data [16]. The assemblies generated using SMRT sequencing have also indicated that a substantial amount of the sequence is missing from the GRCh38 version of the human reference. For example, 12.8 Mb of novel sequences (NS) were detected in the Chinese HX1 assembly [14], and a recent study of 17 individuals from five diverse populations sequenced using linked-read technology revealed 2.1 Mb of NS [18]. Also, an average of 0.7 Mb per individual not present in GRCh38 was found among the 10,000 samples in a population-scale Illumina WGS project [8], showing that NS can, to some extent, also be detected in short read data.

The human de novo assemblies available based on long-read data thus indicate that each personal genome contains a significant amount of dark matter of SV that is not detected by short-read WGS, and several million bases of a NS that cannot be matched to GRCh38. At present, it is unknown how many of these SVs and NS are common to all humans, and thus represent errors in the GRCh38 reference, and how many of them are polymorphic between individuals. To address this question, there is a need to assemble several personal genomes from different populations around the world to a high degree of completeness. Such a collection of de novo genomes would make it possible to improve on GRCh38, and eventually to construct complete new population-specific genomes. In this study, we performed de novo assembly of genomes from two individuals from the Swedish population, in order to investigate the missing pieces of GRCh38, and to evaluate the benefits using a local reference for single nucleotide variant (SNV) calling in population-based WGS data.

2. Materials and Methods

2.1. Samples

Swe1 and Swe2 were selected from the 1000 individuals included in the SweGen project [1]. Samples of whole blood from these two individuals were collected in 2006 and frozen without the separation of white and red blood cells at −70 °C on site, as part of the Northern Sweden Population Health Study (NSPHS), which aims to study the medical consequences of lifestyle and genetics. Genomic DNA was extracted using organic extraction. The NSPHS study was approved by the local ethics committee at the University of Uppsala (Regionala Etikprövningsnämnden, Uppsala, 2005:325 and 2016-03-09). All participants gave their written informed consent to the study, including the examination of environmental and genetic causes of disease, in compliance with the Declaration of Helsinki.

2.2. PacBio Library Preparation and Sequencing

Four PacBio libraries were produced for each of the Swe1 and Swe2 samples using the SMRTbell™ Template Prep Kit 1.0 (PacBio, Menlo Park, CA, USA) according to the manufacturer's instructions. In brief, 10 μg of genomic DNA per library was sheared into 20 kb fragments using the Megaruptor system, followed by exo VII treatment, DNA damage repair, and end-repair before ligation hair-pin adaptors to generate SMRTbell™ libraries for circular consensus sequencing. Libraries were then subjected to exo treatment and PB AMPure bead wash procedures for clean-up before they were size selected with the BluePippin system with a cut-off value of 9500 bp. The libraries were sequenced on the PacBio RSII (PacBio, Menlo Park, CA, USA) instrument using C4 chemistry and P6 polymerase, and a 240 min movie time in a total of 225 SMRTcells™ per sample.

2.3. De Novo Assembly of SMRT Sequencing Reads

Raw data was imported into SMRT Analysis software 2.3.0 (PacBio) and filtered for subreads longer than 500 bp or with a polymerase read quality above 75. A de novo assembly of filtered subreads was generated using FALCON [19] assembler version 0.4.1 (configuration file is provided as Supplementary Information). In order to improve the accuracy of the assembly, two rounds of sequence polishing were performed using the Quiver consensus calling algorithm [19]. For subsequent analysis, primary contigs shorter than 20 kb were excluded from the Swe1 and Swe2 assemblies to reduce putative assembly errors. This is slightly more conservative compared to the Korean AK1 study [15], where a 10 kb cut-off of primary contigs was used.

2.4. Generation of BioNano Optical Maps and Hybrid Assembly

DNA extraction for optical maps was performed at BioNano Genomics (San Diego, CA, USA), starting from frozen blood from Swe1 and Swe2. Optical mapping was performed on the Irys system (BioNano Genomics) using the two labeling enzymes BssSI and BspQI for each individual. The resulting data was used for a two-step hybrid assembly of the PacBio contigs using the IrysView software.

2.5. The hg38 Reference Genome

The hg38 reference genome used in this study is identical to the original, full analysis set of GRCh38 (accession GCA_000001405.15) described in a study by Zheng-Bradley et al. (2017) [20]. This implies that hg38 consists of the primary GRCh38 sequences (autosomes and chromosome X and Y), mitochondria genome, un-localized scaffolds that belong to a chromosome without a definitive location and order, unplaced scaffolds that are in the assembly without a chromosome assignment, the Epstein-Barr virus (EBV) sequence (AJ507799.2), ALT contigs, and the decoy sequences (GCA_000786075.2).

2.6. Quality Control and Alignment of the Two Swedish De Novo Assemblies

Components of MUMmer3 [21] (NUCmer, delta-filter, and dnadiff) were used to assess the quality of the Swe1 and Swe2 de novo assemblies and to perform genome alignments. NUCmer (-maxmatch–l 150 –c 400) was used to align each of the assemblies to hg38. After the alignments, delta-filter (-q) was used to filter out repetitive alignments and to keep the best alignment for each assembled contig. Summary statistics for the filtered whole genome alignments were generated by dnadiff.

2.7. Detection of Structural Variation in PacBio Data

We utilized NGMLR (v0.2.3) (https://github.com/philres/ngmlr) and Sniffles [22] (v1.0.5) to detect SVs from PacBio long reads. Filtered subreads (min subread length 500 bp and polymerase read quality above 75) were first aligned to the hg38 using NGMLR with default parameters. Sniffles (-s 10 –l 50) was subsequently used to identify SVs $\geq$ 50 bp with at least 10 reads support. Only SV detected on chr1-22, X, and Y from Swe1, and on chr1-22 and X from Swe2, were kept for analysis.

2.8. Detection of Novel Sequences

To identify NS, we performed two rounds of sequence mapping using NUCmer [21]. The first round of mapping was the same as described above, when aligning contigs to the hg38 reference. Contigs and part of contigs that failed to align to the reference in the first step were then processed in a second round of mapping, where more relaxed settings were used in an attempt to have more sequences aligned by NUCmer (-maxmatch–l 100–c 200). Duplicated sequences were then removed to obtain a set of NS for each of the two individuals (Swe1 and Swe2). All NS in the final set have a sequence length of at least 100 bp, with a sequence identity to the hg38 reference that is less than 80%.

2.9. Repeat Analysis and BLAST Comparison of Novel Sequences

Repeats in NS were analyzed using RepeatMasker (-species human -s–x; http://www.repeatmasker. org). NS were searched against the nucleotide collection database using BLAST [23] (2.2.31+) (-max_target_seqs 1–task blastn–num_threads 16). In order to obtain matched sequences of a relatively high similarity, the BLAST results were post processed by setting an E-value threshold at 10^{-50} and by keeping only the top hit for each NS.

2.10. Anchoring Novel Sequences on Human Chromosomes

To determine the potential genomic position of the NS that may be anchored, we first used NUCmer (-maxmatch–l 100–c 200) and delta-filter (-q) to map the NS to the hybrid scaffolds that were generated by PacBio contigs and BioNano optical maps. After anchoring of the hybrid scaffolds, NUCmer (-maxmatch–l 150–c 400) and delta-filter (-q) were ran to identify the location of the alignments on hg38 chromosomes. NS that mapped to anchored hybrid scaffolds were further analyzed to identify unique or multiple location anchors. NS that were anchored to decoy sequences included in the hg38 reference were excluded from the final results.

2.11. Construction of an Extended Reference Based on Swedish Novel Sequences

Novel sequences detected in Swe1 and Swe2 were appended to hg38 to create an extended version of the human reference sequence (named hg38+NS). For NS overlapping between both Swedish individuals, only the Swe1 version of the NS was used. The resulting hg38+NS reference added 17.3 Mb of NS to hg38.

2.12. Re-Alignment of SweGen Illumina Data to hg38 and hg38+NS

In total, 200 of the SweGen samples [1] were processed with the Cancer Analysis Workflow (CAW) pipeline (https://github.com/SciLifeLab/CAW) in normal-only mode (no tumor samples), once with hg38 and again with hg38+NS as the reference. CAW implements a workflow based on GATK best practices. In summary, reads were aligned using BWA-MEM 0.7.15 with the ALT-aware option turned off. Duplicates were then marked with Picard's MarkDuplicates 2.0.1. The tools RealignerTargetCreater, IndelRealigner, CreateRecalibrationTable, HaplotypeCaller, and GenotypeGVCFs from GATK 3.7.0 (https://software.broadinstitute.org/gatk/) were then used in that order to realign around indels, recalibrate base qualities, and call variants, respectively, resulting in a final CRAM and VCF file for each sample.

Also, a similar analysis was re-run for 150 of the 200 SweGen samples, but with the ALT aware option turned on in the BWA-MEM alignment. This ALT aware analysis resulted in a much higher number of lost and gained SNVs compared to the non-ALT aware alignment. We therefore decided to focus on the non-ALT aware analysis in this study, i.e., the analysis run on the 200 samples. By a non-ALT aware alignment, we get a conservative estimate of the number of lost and gained SNVs and do not exaggerate the effect of adding NS to the hg38 reference.

2.13. Analysis and Annotation of SNVs in SweGen Re-Alignments

To detect SNVs that were consistently gained or lost among the 200 SweGen samples when the NS were added to hg38, we employed a filtering strategy using custom scripts in Perl and R. ANNOVAR [24] was used to annotate gained and lost SNVs with information about human genetic variation from dbSNP [25] v147 and protein coding genes from the NCBI RefSeq database [26].

3. Results

3.1. De Novo Assembly of Two Swedish Individuals

To construct two high-quality genome references for the Swedish population, DNA was extracted from blood samples obtained from one male (Swe1) and one female (Swe2). The two individuals were unrelated and selected from the 1000 samples included in SweGen, which is a project where the genetic variation in a cross-section of the Swedish population was studied using Illumina WGS [1]. A principal component analysis (PCA) shows that Swe1 and Swe2 are relatively distant from each other in the context of the genetic variation within Sweden (Figure 1A). The long tail of SweGen samples that are intermixed with the Finnish genomes mainly represents genome sequences from the northern parts of the country, and thus the two genomes contain a large portion of the common genetic variation in the Swedish population.

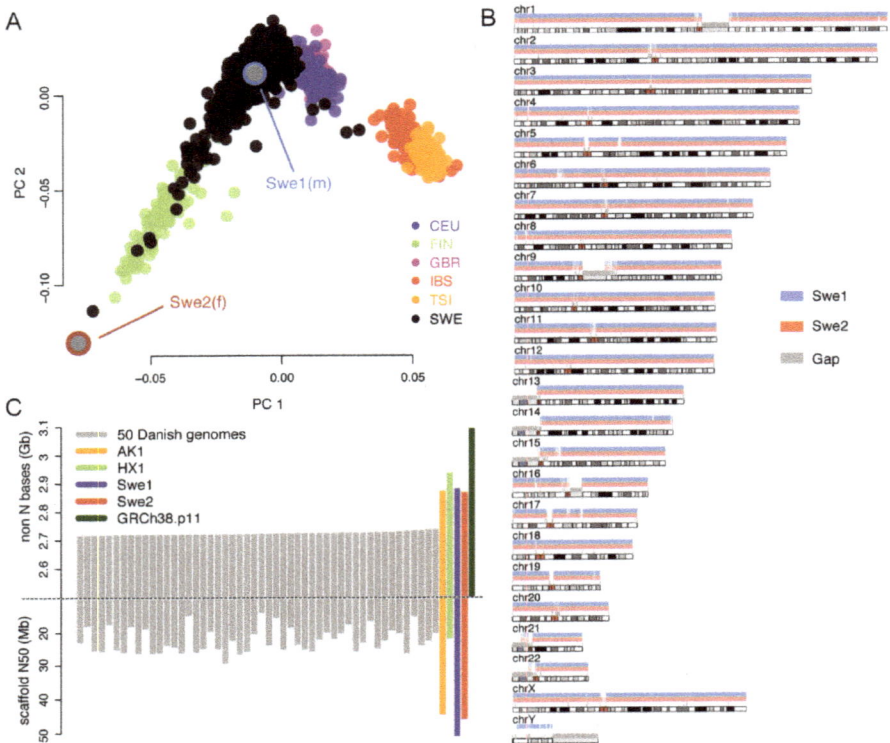

Figure 1. Selection of individuals and de novo assembly results. (**A**) Results of principal component analysis (PCA) of whole genome sequencing (WGS) data from the SweGen project [1], compared to the European 1000 Genomes data [27] (CEU: Utah Residents with Northern and Western Ancestry, FIN: Finnish in Finland, GBR: British in England and Scotland, IBS: Iberian Population in Spain, TSI: Toscani in Italia). The black dots indicate 942 samples from the Swedish Twin Registry (STR), which were sequenced within the SweGen project and represent a cross-section of the Swedish population. Swe1 and Swe2 are the individuals selected for de novo sequencing. (**B**) Alignment of contigs for Swe1 (blue) and Swe2 (red) to the human GRCh38 reference. A total of 6812 contigs could be aligned for Swe1 and 6924 for Swe2. Only the male Swe1 sample has extensive coverage of the Y chromosome. (**C**) The bars show the total number of non-N bases (top) and scaffold N50 values (bottom) for Swe1, Swe2, and a selection of other human de novo assemblies. The grey bars represent the top 50 genomes with the highest number of non-N bases from an Illumina mate-pair assembly of 150 individuals [10]. The Korean (AK1) and Chinese (HX1) genomes were assembled by a combination of single-molecule real-time (SMRT) sequencing and optical mapping. Scaffold N50 is not shown for GRCh38 (in green) since it is much higher than for the personal genomes and difficult to fit into the same plot.

SMRT sequencing data was generated at an average coverage of 78.7× for Swe1 and 77.8× for Swe2 (Table S1). By de novo assembly [28], followed by two iterations of genome polishing, we were able to construct sequence assemblies of 2.996 Gb and 2.978 Gb for Swe1 and Swe2, consisting of 7166 and 7186 contigs, respectively (Table S2). Each of the assemblies contained about 3000 primary contigs and an additional 4000 alternative contigs originating from regions with high heterozygosity. The alternative contigs only cover a small fraction of the genome; about 115 Mb in each individual. N50 values for the primary contigs were 9.5 Mb for Swe1 and 8.5 Mb for Swe2. For both individuals, we also generated BioNano optical mapping data with two different labeling enzymes, at over 100× coverage per enzyme. A two-step hybrid scaffolding of the SMRT sequencing contigs together with the optical maps resulted in assemblies of size 3.1 Gb and scaffold N50 of 49.8 Mb (Swe1) and 45.4 Mb (Swe2) (Table S3). These numbers are similar to the 44.8 Mb scaffold N50 obtained for the first published Korean genome [15], and substantially larger than the median scaffold N50 of 21 Mb obtained for 150 Danish genomes [10]. It is worth noting that the DNA samples used for optical mapping of Swe1 and Swe2 were extracted from blood collected in 2006. Our results thus demonstrate that it is possible to obtain very high-quality genome assemblies starting from frozen blood that has been stored in the freezer for over a decade.

3.2. Evaluating the Quality of the De Novo Assemblies

To assess the quality of the two de novo assemblies, we aligned the contigs for Swe1 and Swe2 to the hg38 reference genome. Throughout this article, the abbreviation hg38 is used to denote a sequence that is identical to the full analysis set of GRCh38, which includes un-localized scaffolds and decoy sequences [20] (see Methods). For Swe1 and Swe2, respectively, 2.971 Gb (99.14%) and 2.956 Gb (99.24%) of the assembled sequence could be uniquely aligned to hg38 (see Figure 1B and Table S4). The slightly higher number of aligned bases for Swe1, who is a male, can be explained by sequences on the Y chromosome that are not present in the female Swe2 sample. The average identity between the contigs and hg38 was over 99.7% for both genomes. Intriguingly, a higher fraction of the Swe2 sequence data can be uniquely aligned to the Swe1 de novo assembly (99.55%) compared to hg38 (99.24%), thus suggesting that the hg38 reference does not contain all sequences present in these Swedish individuals. The corresponding analysis for Swe1 is not relevant in this context, since Swe1 is expected to contain a sequence on the Y chromosome not present in the female Swe2.

In order to discover a NS missing from hg38, it is essential that Swe1 and Swe2 were assembled to a high degree of completeness and with as few gaps as possible. To evaluate this, we compared our Swedish de novo assemblies to results obtained for the Korean AK1 [15], the Chinese HX1 [14], and 150 Danish genomes [10]. As seen in Figure 1C, the primary contigs of Swe1 and Swe2 contain a similar number of unambiguous (non-N) bases as the other SMRT sequencing assemblies (i.e., AK1 and HX1). Importantly, the assemblies obtained from SMRT sequencing contain over 100 Mb of additional sequence as compared to the Illumina mate-pair assemblies. The GRCh38 reference contains almost 3.1 Gb, which is significantly more compared to the ~2.9 Gb for Swe1 and Swe2. To a certain extent, these differences can be explained by the fact that GRCh38 is based on a combination of DNA sequences and haplotypes from several individuals [9], which could lead to an inflated genome size, and also that primary contigs shorter than 20 kb were excluded from the Swe1 and Swe2 assemblies. N50 scaffold values are highest for the Swe1, Swe2, and AK1 assemblies, which all used BioNano data from two labeling enzymes for hybrid scaffolding. A single enzyme was used for HX1 and this assembly has a scaffold N50 similar to those obtained for the Danish genomes.

3.3. Structural Variation in Swedish Genomes

Analysis of SV resulted in a total of 17,936 SVs for Swe1 and 17,687 SVs for Swe2 (Table S5). These numbers can be compared with the 20,175 and 18,210 SVs detected in the Chinese HX1 and in the Korean AK1 assembly, respectively. The SV length distribution shows an enrichment of ALU repeat elements at around 300 bp and of LINE elements at around 6100 bp, similar to what has been previously reported [15] (Figure S1).

3.4. Detection of Novel Sequences Not Present in the Human Reference

Even though most of the contigs in our assemblies were in good agreement with the human reference, we detected 25.6 Mb of sequences in Swe1 and 22.6 Mb in Swe2 that could not be aligned to hg38 (Table S4). To refine these sequences further, we performed a two-step re-alignment using more relaxed settings and removed duplicated sequences (see Methods). This resulted in 2859 NS in Swe1 of a total length of 13.8 Mb, and 2786 NS in Swe2 of a total length of 10.6 Mb (Table S6 and Data S1). The NS were required to be at least 100 bp in length and have at most 80% identity to hg38. They could either originate from contigs that could not be aligned to hg38, or from inserted elements in the aligned contigs. As seen in Figure 2A, most of the NS are relatively short (between 100 bp and 5 kb). However, 83% of NS bases originate from sequences that are over 5 kb in length.

Repeat masking using sensitive settings, revealed an abundance of repetitive elements in the NS (see Figure 2B and Table S7). For Swe1, 88.58% of the NS bases were found to be repetitive. A slightly lower repeat content, 83.60%, was detected in the NS from Swe2. Since the repeat content is around 50% among all Swe1 and Swe2 contigs, there is a high enrichment of repeats in the NS. Also, the GC level is slightly elevated, with values of 42.68% (Swe1 NS) and 43.45% (Swe2 NS), compared to 40.95% in all contigs (Table S8). SMRT sequencing is known to perform well in repetitive regions and high-GC regions, and therefore these results are not unexpected. Annotation of all the repeats showed that satellites and simple repeats make up 82% of the NS bases, but only 3% of all of the primary contig sequences (see Table S7). Interestingly, all other groups of repeats are underrepresented among the NS. Even though a high proportion of the NS are repetitive, there is also a substantial amount of non-repetitive sequence. For Swe1 and Swe2, 1.58 Mb and 1.73 Mb of NS remained after the repeat masking.

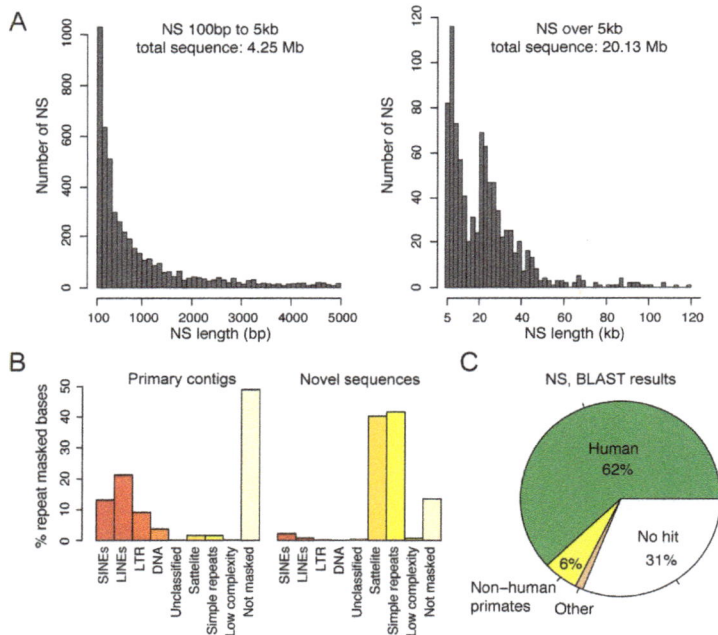

Figure 2. Characterization of novel sequences (NS) found in Swe1 and Swe2. (**A**) The histograms show the length distribution of all NS found in Swe1 and Swe2. Shorter NS are displayed in the left panel (100 bp to 5 kb), and longer NS are shown in the right panel (>5 kb). The longer NS comprise the majority of the NS in Swe1 and Swe2. (**B**) Results of repeat masking in primary contigs (left) and NS (right). Within the primary contigs, 51% of the bases are found to be repetitive using the repeat masker software, while 86% of the bases are repetitive within the NS. Satellite and simple repeats make up 82% of the bases in the NS. (**C**) Results of matching the 5645 NS in Swe1 and Swe2 to the NCBI database using BLAST [23]. Each piece of the pie chart represents the number of NS that were assigned to a particular species as the top hit. The No hit category (in white) contains NS where no E-value reached 10^{-50} or lower. A total of 72 of the NS are in the Other category, which includes matches to a number of parasitic worms (both for Swe1 and Swe2) and a complete human papilloma virus 35 (HPV35) genome (only for Swe2).

3.5. Origin of the Novel Sequences

To further investigate the contents of the NS, we performed a BLAST search [29] against all sequences present in the NCBI database (see Figure 2C, Table S9 and Data S2). Thirty-one percent of the NS did not produce a BLAST hit, implying that they have not been previously reported. For the remaining NS, the majority matched to human entries in NCBI, thereby suggesting that a majority of our NS have been detected previously, but originate from regions or haplotypes that have not been included in the hg38 reference. We also detected 5% non-human primate sequences, which most likely originate from regions missing in hg38 that have been sequenced in another primate. Nearly 1% of the NS match to other, non-primate, species. Of note, several of the hits show high similarity to parasitic worms, including *Spirometra erinaceieuropaei*, *Enterobius vermicularis*, and *Dracunculus medinensis*. Since it is highly unlikely that the two Swedish individuals indeed have DNA from these parasites present in their blood, a more plausible explanation is that the worm genome assemblies contain a fraction of human sequence. The initial worm assemblies were based on short-read sequencing of samples extracted from human patients [30], and contigs not aligning to GRCh38 may have been mis-annotated as the worm sequence, thus explaining the overlap with our NS. Notably, the Swe2 sample also contained a complete human papilloma virus 35 (HPV35). This could either originate from an HPV35 infection in the blood of this individual, or from a contamination in the sample.

3.6. Comparing Novel Sequences between Swedish Individuals and the Chinese HX1

To investigate whether the NS are individual-specific, population-specific, or shared between different populations, we compared our results to those obtained for the Chinese HX1 genome [14]. The HX1 assembly was based on 103× genome-wide SMRT sequencing of DNA from a human blood sample, where 12.8 Mb of NS was found. Starting from the NS identified in Swe1 or Swe2, we determined whether the same sequences could be identified in the other Swedish assembly, or in HX1 (see Figure 3A and Table S10). For Swe1 and Swe2, 55% (7.65 Mb) and 52% (5.51 Mb) of the NS, respectively, could also be found in the other Swedish individual, as well as in HX1. The higher overlap obtained when starting the analysis from Swe1 is explained by certain repetitive elements that occur with a higher copy number in Swe1 compared to Swe2. Our results also show the presence of over 5 Mb of NS in the three-way overlap category, i.e., found in all three individuals. A smaller amount of NS (~1.5 Mb) was only common between the two Swedish individuals, while not found in the Chinese HX1, thus representing a possible population-specific sequence. We also identified a substantial amount of individual-specific sequences, 3.27 Mb for Swe1 and 3.22 Mb for Swe2. Interestingly, a much higher amount of NS were shared between Swe1 and HX1 (1.36 Mb) compared to Swe2 and HX1 (0.29 Mb). Since Swe1 and HX1 are both males, while Swe2 is a female, the ~1 Mb of additional NS shared between Swe1 and HX1 may at least partly be explained by segments of the Y chromosome that are missing from the hg38 reference.

3.7. Anchoring Novel Sequences on Human Chromosomes

We next aimed to anchor the NS onto human chromosomes using information provided by the PacBio long-read data and BioNano optical maps (see Methods section). Only a minority of sequences could be placed into the human genome using this approach, but this analysis still provided valuable insights about the genomic localization of the NS (see Figure 3B). For Swe1 and Swe2, 2.08 Mb and 1.97 Mb of NS, respectively, could be uniquely anchored to a chromosome, while 1.70 Mb and 1.55 Mb were anchored to multiple chromosomes (see Table S11). This again shows that many NS contain repetitive or transposable elements. For the uniquely anchored NS, we observed an accumulation at certain chromosomes. The highest amount of three-way overlap sequences is present on chromosome 21, while chromosomes 13, 14, and 22 also show enrichment (Figure 3C). The NS placed on these chromosomes are mainly localized to centromeric or telomeric regions, suggesting that placement of these sequences has previously been difficult to determine due to their repetitive content. Interestingly, we detected an accumulation of population-specific NS present in both Swedish individuals, but not in the Chinese HX1 on chromosome 17. A relatively large amount of sequences (121 kb) shared only between the two male individuals (Swe1 and HX1) could be anchored to the Y chromosome. Surprisingly, we also noted an accumulation of NS shared between Swe1 and HX1 on chromosome 17.

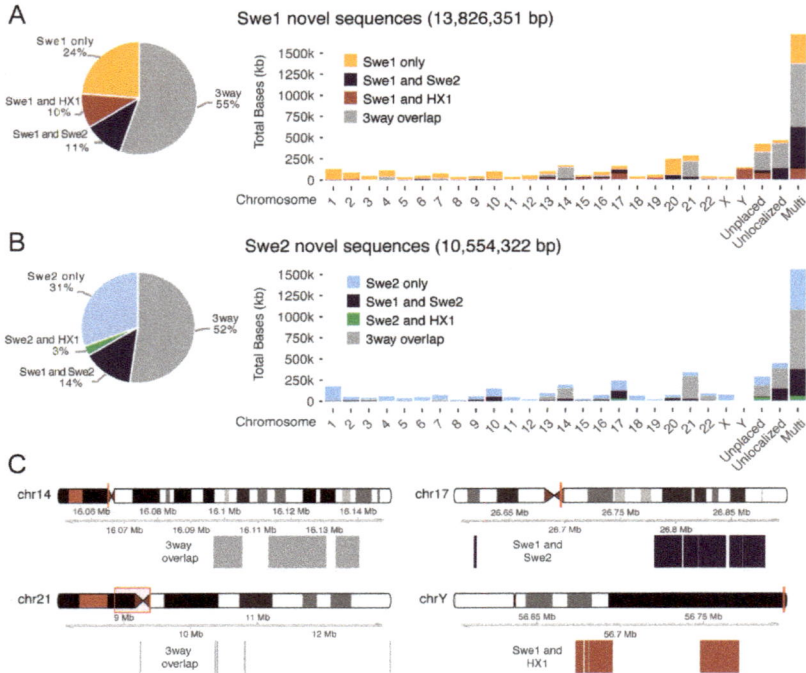

Figure 3. Anchoring of Swe1 and Swe2 NS to the hg38 reference. (**A**) The pie chart to the left shows the proportion of Swe1 NS (in total 13.8 Mb) that are also found in Swe2 or in the Chinese HX1. The category 3way (grey) represents NS that are found in all three individuals. The bars to the right show the amount of NS that can be anchored to the hg38 genome. The category unplaced represents sequences in hg38 that are not associated with any chromosome, and unlocalized corresponds to sequences that are associated with a specific chromosome but have not been assigned an orientation and position. The multi category furthest to the right represents NS that are mapping to multiple chromosomes. (**B**) Similar results for NS detected in Swe2. (**C**) Examples of chromosomal regions where a high amount of NS are detected. The two plots to the left show the localization of 3way overlap sequences (i.e., found in Swe1, Swe2, and HX1) near the centromeric regions of chr14 and chr21. The top right panel displays a region on chr17 where an excess of NS found only in Swe1 and Swe2 could be anchored. The bottom left panel shows NS detected only in the two males (Swe1 and HX1) that could be anchored to regions close to the telomere of chromosome Y.

3.8. Application of Novel Sequences for Population Scale WGS Analysis

Having identified several Mb of DNA not present in the human reference, we were interested to see whether these NS would improve the results of whole genome re-sequencing of the Swedish population. We therefore created a new reference consisting of hg38 combined with all the NS detected in Swe1 and Swe2 (named hg38+NS), after which we leveraged the Illumina WGS data from the SweGen dataset [1] and aligned the reads from 200 individuals both to hg38, as well as to hg38+NS. The aim of this analysis was to study whether the number of SNVs was altered as a result of appending NS to the reference, through an analysis procedure outlined in Figure 4A.

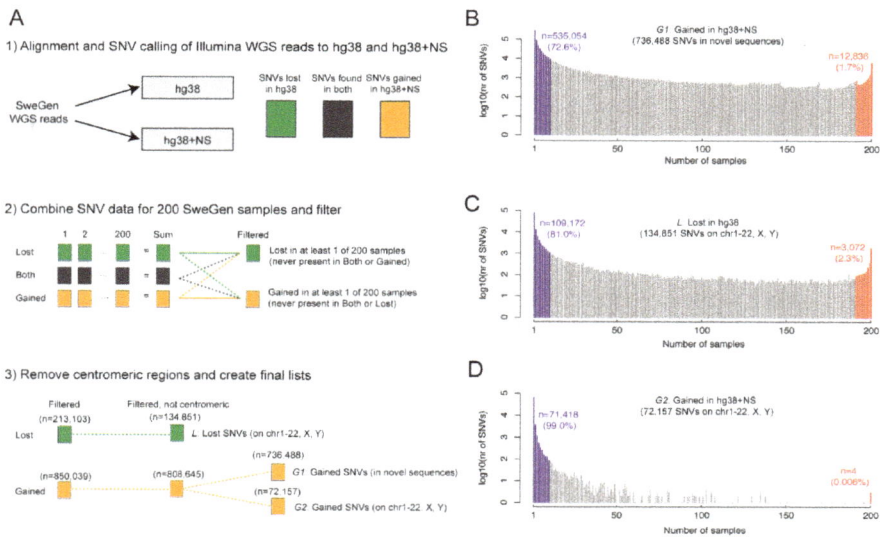

Figure 4. Re-analysis of Illumina WGS data using a Swedish human reference. (**A**) Overview of our method to evaluate the effect of NS on SNV calls from Swedish Illumina WGS data. In the first step, reads from 200 SweGen samples [1] were aligned both to hg38 and to an extended reference (hg38+NS), where 17.3 Mb of NS detected in Swe1 and Swe2 were appended to hg38. In step 2, single nucleotide variants (SNVs) for each of the samples were sorted into three groups: (i) SNVs found only in hg38, but not in hg38+NS (named Lost, in green); (ii) SNVs found both in hg38 and hg38+NS ('Both', in grey); and (iii) SNVs found only in hg38+NS, but not in hg38 (Gained, in orange). After such SNV tables were generated for all 200 individuals, a summary file was created for the Lost and Gained group. The Lost SNVs were not allowed to be detected in any of the Gained or Both files. A similar filtering was also performed for the Gained group. In step 3, we further filtered the SNV lists by removing all centromeric regions (from file centromeres_USCS_hg38.txt). The resulting Gained SNVs were separated into two distinct groups, those present in hg38 chromosomes (chr1-22, X or Y) and those present in the NS. (**B**) Frequency distribution of the 736,488 SNVs that were gained in the NS. The x-axis shows the not

21 SweGen samples (out of 200) and the *y*-axis show the number of gained SNVs for each number of samples on a log10-scale. Most of the gained SNVs are detected only in a few samples. The blue and red areas show the number of SNVs that are gained in at most 5% and at least 95% of samples, respectively. (**C**) Frequency distribution of the SNVs that were lost in hg38 when adding NS to the hg38 reference. (**D**) Frequency distribution of the gained SNVs on chromosomes 1-22, X, or Y (i.e., not in NS) when adding NS to the hg38 reference.

An average of 42.5 SNVs per kb was detected in the NS and their frequency distribution is shown in Figure 4B. Some of these SNVs are likely to represent true novel genetic variation in the cohort, but a fraction may also originate from errors in the Swe1 and Swe2 assemblies. Surprisingly, the addition of NS to the reference had a large effect on variant calls of the autosomes and sex chromosomes, where 134,851 SNVs disappeared (outside of centromeric regions) when the extended reference was used (Figure 4C). These SNVs originate from reads that preferentially align to a NS and can be considered as false positives in hg38. Interestingly, we also found a substantial number of SNVs ($n = 72,157$) which were gained on the hg38 chromosomes when using the extended reference. These gained SNVs in hg38 have overall lower allele frequencies compared to the lost SNVs (see Figure 4D).

Finally, we investigated SNVs that were consistently lost or gained in hg38 for at least 5% of the 200 SweGen samples when using the extended reference (see Figure 5A). Only a small number of SNVs (*n* = 823) were gained on the hg38 chromosomes in at least 5% of the samples when using the extended reference. However, 26,724 SNVs were lost in at least 5% samples when appending NS to the hg38 reference. These consistently lost SNVs have an uneven distribution over the genome, with the highest peak on chrY and smaller peaks on several other chromosomes. Global annotation of the consistently lost SNVs showed that 7130 (27%) of these are present in version 147 of dbSNP. For the consistently gained SNVs, only 130 (16%) are present in dbSNP, suggesting that these SNVs are more difficult to detect using the hg38 reference alone. A total of 109 consistently lost SNVs were located in a coding sequence of a gene, but none of the consistently gained SNVs were in coding regions. Figure 5B shows an example region on chr17 where the NS improved the alignment of Illumina WGS data for two SweGen individuals, resulting in the removal of around 100 false positive SNVs, and importantly, the discovery of seven novel SNVs that were previously masked by the mis-aligned reads. A second example is shown in Figure 5C where a region on chrY with about 1000× coverage and many dubious SNVs are cleaned up when NS are appended to hg38. In a third example, as illustrated by the genome browser view of the *FRG2C* locus, the hg38+NS reference improves alignments in coding regions (see Figure 5D).

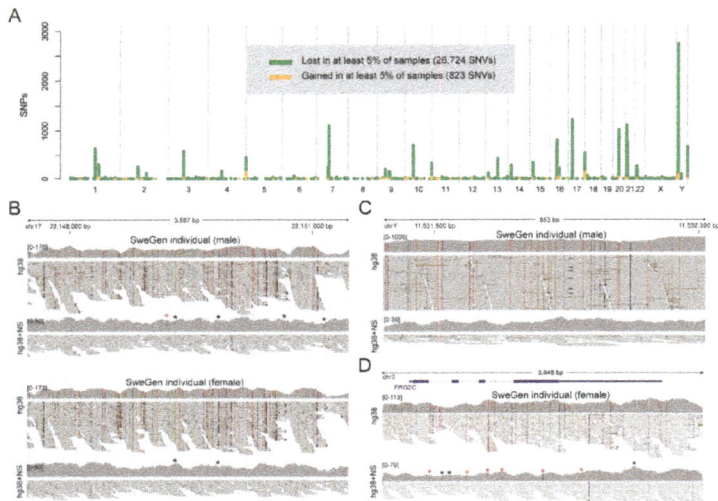

Figure 5. A novel reference gives improved alignment and SNV calling of SweGen WGS data. (**A**) Genomic distribution of SNVs that are lost (green) and gained (orange) when NS are appended to the hg38 reference. Only non-centromeric SNVs that are lost/gained in at least 5% of the 200 SweGen samples are shown in this figure. (**B**) An IGV [31] view of Illumina reads for two representative SweGen samples at a region on chr17, where some SNVs are lost and others are gained when using the hg38+NS reference. Illumina data is shown for a male and a female (not the same individuals as Swe1 and Swe2). Both for the male and female, the coverage decreases over the region when NS are appended to hg38, and about 100 (homozygous) false positive SNV calls are lost in each of the samples. Only five heterozygous SNVs where found for the male individual when the novel reference was used, and two homozygous SNVs for the female (marked by asterisks '*'). A red asterisk indicates a gained SNV that is not detected in hg38. (**C**) An example region on chrY where the coverage was reduced from almost 1000× to below 30× when using hg38+NS, and where a large number of SNVs were lost. Only data for the male individual is shown in this panel. (**D**) Improved alignment and SNV calling over the *FRG2C* locus on chromosome 3. A large number of SNVs were lost, and six SNVs were gained (red asterisks '*'), in the female SweGen sample. Some of the lost and gained SNVs are located in the coding sequences of *FRG2C*.

4. Discussion

Swe1 and Swe2 represent two of the most complete individual human de novo assemblies produced to date. On average, the primary contigs contain 2.87 Gb of unambiguous (non-N) sequence per individual, which is similar to the Chinese HX1 [14] and the Korean AK1 [15] individuals, and 133 million more bases than what could maximally be assembled in any of the 150 Danish genomes [10]. Even though our Swedish assemblies are comparable to HX1 and AK1, both in terms of quality and completeness, they are still not entirely complete. To a large extent, this incompleteness can be explained by the current limitations of the sequencing and optical mapping technologies used in this study. Not even the longest DNA molecules in our data can bridge repeats that are spanning over several mega bases, and this implies that our assemblies will break down at such loci. Eventually, the Swe1 and Swe2 assemblies could be further improved by BAC sequencing of specific regions [15], by chromosome interaction mapping [32], by linked-reads from 10× Genomics [15,17], or long reads from nanopore-based sequencing [33]. By combining data from these technologies, it would also be possible to phase haplotypes, and to generate diploid sequences over large parts of the Swe1 and Swe2 genomes.

With the new sequencing technologies, it is possible to assemble complete human genomes starting directly from tissues or blood, instead of using cell line samples that have been the traditional source of DNA for human reference genomes. This is of importance since it should resolve the potential issue with genomic aberrations introduced during cell line transformation and long-term culturing [34]. In this study, we have therefore chosen to compare the assemblies of Swe1 and Swe2 to the Chinese HX1 genome, which was also obtained from a blood sample, rather than the Korean AK1 genome, which was based on cell line DNA.

The Swedish de novo assemblies reveal a large amount of NS not present in the GRCh38 reference. Over 5 Mb of NS were overlapping between our two Swedish genomes and the Chinese HX1, and this likely represents a valid human DNA sequence. In addition, 1.36 Mb of male-specific sequence was found (i.e., overlapping between Swe1 and HX1) and a substantial fraction of these sequences could be anchored to the Y chromosome. We estimate the total amount of sequence missing from GRCh38 to be at least 6 Mb, which corresponds to about 0.2% of the size of the human genome. Another interesting class of NS are the ~1.5 Mb found in Swe1 and Swe2, but not in HX1. Several of these potentially population-specific NS are clustered at certain genomic regions, such as chr17 (see Figure 3C). Some of these regions could have been targets for selection during human evolution, although this needs to be investigated further.

At this point, there is no evidence for the presence of functional elements within the NS, although preliminary data suggest that some of them are actively transcribed (data not shown). A more thorough analysis would be required to shed light on the functional relevance of these NS in the different cell types in the human body. For example, this could involve searching for open reading frames, conserved sequences, expressed mRNAs, enhancers, and transcription factor binding motifs. It might even be possible to leverage mass spectrometry data to search for peptide sequences missing from the current version of the human genome. Although the functional analysis is highly relevant, it would be a major undertaking that falls outside the scope of this present study.

Our results show that the NS can be used to construct a new version of the human genome reference that improves the analysis of population-scale Illumina WGS data. On average, 10,898 SNVs per individual were lost, and 75,035 SNVs per individual were gained in 200 SweGen samples when appending the NS to GRCh38, with some of this variation affecting the coding sequences of known genes. Because of the stringent filtering options used in our analysis (see Figure 4A), these numbers should be seen as a conservative estimate of the novel variation that could be resolved using an improved reference. In addition, since many regions still show poor alignments for SweGen data also when using hg38+NS (data not shown), it is likely that our reference could be further improved and customized for the Swedish population. This could for example be done by flipping genetic variants so that the common alleles in the Swedish population are represented in the reference sequence, as this is an approach that has been suggested to improve the alignments of population specific NGS data [35]. However, the benefits of an improved reference are likely to be even stronger for other, non-European, population groups that were poorly represented in the original assembly of GRCh38 [9].

5. Conclusions

In conclusion, despite all efforts to refine the human genome since its original release in 2001 [36], our results indicate that substantial improvements could still be made, not least to represent specific population groups, by the de novo assembly of representative human genomes from different populations.

Supplementary Materials: The following are available online at http://www.mdpi.com/2073-4425/9/10/486/s1, The Swe1 and Swe2 raw sequence data and assembly files will be made available during 2018 from a local Swedish installation of the European Genome-phenome Archive (EGA) (https://www.ebi.ac.uk/ega) that is now being implemented at Uppsala University and SciLifeLab. The dataset has the following doi:10.17044/NBIS/G000006. In addition, the following data files are made available as supplementary material: Supplementary Data S1: Novel sequences in Swe1 and Swe2; Supplementary Data S2: BLAST hits in novel sequences. Figure S1. Length distribution of structural variants detected in Swe1 and Swe2. (A) Lengths of insertions (red) and deletions (blue) in Swe1 and Swe2 ranging from 50 bp to 1 kb. (A) Lengths of insertions (red) and deletions (blue) in Swe1 and Swe2 ranging from 1 kb to 10 kb. Table S1. SMRT-sequencing data overview. Table S2. Results of FALCON *de novo* assembly. Table S3. Overview of hybrid scaffolding of PacBio data using BioNano optical maps. Table S4. Alignment results of PacBio data to hg38. Table S5. Structural variation results for Swe1, Swe2 and HX1. Table S6. Statistics for NS in Swe1 and Swe2. Table S7. Repeat contents for NS is Swe1 and Swe2. Table S8. GC contents for NS is Swe1 and Swe2. Table S9. BLAST results for NS. Table S10. Overlap of NS between Swe1, Swe2 and HX1. Table S11. Amount of NSs that could be anchored to hg38.

Author Contributions: A.A. and U.G. conceived the study. H.C., M.M., P.O., I.B., J.D., S.H., F.V., J.N. and A.A. analyzed the data. I.H. and S.H. optimized and performed wet lab experiments. A.A., L.F. and U.G. wrote the manuscript with input from all authors. All authors read and approved the final manuscript.

Funding: This work was funded by Science for Life Laboratory (SciLifeLab) as a National Project, supported by the Knut and Alice Wallenberg Foundation (2014.0272), and the Swedish Research Council (PI:UG).

Acknowledgments: PacBio SMRT sequencing was performed by the National Genomics Infrastructure (NGI), which is hosted by SciLifeLab in Uppsala. Optical maps were generated at BioNano (CA, USA). Computations were performed on resources provided by SNIC through Uppsala Multidisciplinary Center for Advanced Computational Science (UPPMAX) under projects b2015225 and sens-2016003. MM and PO were financially supported by the Knut and Alice Wallenberg Foundation as part of the National Bioinformatics Infrastructure Sweden at SciLifeLab.

Conflicts of Interest: The authors declare that they have no competing interests

References

1. Ameur, A.; Dahlberg, J.; Olason, P.; Vezzi, F.; Karlsson, R.; Martin, M.; Viklund, J.; Kahari, A.K.; Lundin, P.; Che, H.; et al. SweGen: A whole-genome data resource of genetic variability in a cross-section of the Swedish population. *Eur. J. Hum. Genet.* **2017**, *25*, 1253–1260. [CrossRef] [PubMed]

2. Boomsma, D.I.; Wijmenga, C.; Slagboom, E.P.; Swertz, M.A.; Karssen, L.C.; Abdellaoui, A.; Ye, K.; Guryev, V.; Vermaat, M.; van Dijk, F.; et al. The Genome of the Netherlands: Design, and project goals. *Eur. J. Hum. Genet.* **2014**, *22*, 221–227. [CrossRef] [PubMed]

3. Fakhro, K.A.; Staudt, M.R.; Ramstetter, M.D.; Robay, A.; Malek, J.A.; Badii, R.; Al-Marri, A.A.; Abi Khalil, C.; Al-Shakaki, A.; Chidiac, O.; et al. The Qatar genome: A population-specific tool for precision medicine in the Middle East. *Hum. Genome Var.* **2016**, *3*, 16016. [CrossRef] [PubMed]

4. Gudbjartsson, D.F.; Helgason, H.; Gudjonsson, S.A.; Zink, F.; Oddson, A.; Gylfason, A.; Besenbacher, S.; Magnusson, G.; Halldorsson, B.V.; Hjartarson, E.; et al. Large-scale whole-genome sequencing of the Icelandic population. *Nat. Genet.* **2015**, *47*, 435–444. [CrossRef] [PubMed]

5. Nakatsuka, N.; Moorjani, P.; Rai, N.; Sarkar, B.; Tandon, A.; Patterson, N.; Bhavani, G.S.; Girisha, K.M.; Mustak, M.S.; Srinivasan, S.; et al. The promise of discovering population-specific disease-associated genes in South Asia. *Nat. Genet.* **2017**, *49*, 1403. [CrossRef] [PubMed]

6. Wong, L.P.; Ong, R.T.; Poh, W.T.; Liu, X.; Chen, P.; Li, R.; Lam, K.K.; Pillai, N.E.; Sim, K.S.; Xu, H.; et al. Deep whole-genome sequencing of 100 southeast Asian Malays. *Am. J. Hum. Genet.* **2013**, *92*, 52–66. [CrossRef] [PubMed]
7. Consortium, U.K.; Walter, K.; Min, J.L.; Huang, J.; Crooks, L.; Memari, Y.; McCarthy, S.; Perry, J.R.; Xu, C.; Futema, M.; et al. The UK10K project identifies rare variants in health and disease. *Nature* **2015**, *526*, 82–90. [CrossRef] [PubMed]
8. Telenti, A.; Pierce, L.C.; Biggs, W.H.; di Iulio, J.; Wong, E.H.; Fabani, M.M.; Kirkness, E.F.; Moustafa, A.; Shah, N.; Xie, C.; et al. Deep sequencing of 10,000 human genomes. *Proc. Natl. Acad. Sci. USA* **2016**, *113*, 11901–11906. [CrossRef] [PubMed]
9. Schneider, V.A.; Graves-Lindsay, T.; Howe, K.; Bouk, N.; Chen, H.C.; Kitts, P.A.; Murphy, T.D.; Pruitt, K.D.; Thibaud-Nissen, F.; Albracht, D.; et al. Evaluation of GRCh38 and de novo haploid genome assemblies demonstrates the enduring quality of the reference assembly. *Genome Res.* **2017**, *27*, 849–864. [CrossRef] [PubMed]
10. Maretty, L.; Jensen, J.M.; Petersen, B.; Sibbesen, J.A.; Liu, S.; Villesen, P.; Skov, L.; Belling, K.; Theil Have, C.; Izarzugaza, J.M.; et al. Sequencing and de novo assembly of 150 genomes from Denmark as a population reference. *Nature* **2017**, *548*, 87–91. [CrossRef] [PubMed]
11. Ross, M.G.; Russ, C.; Costello, M.; Hollinger, A.; Lennon, N.J.; Hegarty, R.; Nusbaum, C.; Jaffe, D.B. Characterizing and measuring bias in sequence data. *Genome Biol.* **2013**, *14*, R51. [CrossRef] [PubMed]
12. Ameur, A.; Kloosterman, W.P.; Hestand, M.S. Single-molecule sequencing: Towards clinical applications. *Trends Biotechnol.* 2018. [CrossRef] [PubMed]
13. Chaisson, M.J.; Huddleston, J.; Dennis, M.Y.; Sudmant, P.H.; Malig, M.; Hormozdiari, F.; Antonacci, F.; Surti, U.; Sandstrom, R.; Boitano, M.; et al. Resolving the complexity of the human genome using single-molecule sequencing. *Nature* **2015**, *517*, 608–611. [CrossRef] [PubMed]
14. Shi, L.; Guo, Y.; Dong, C.; Huddleston, J.; Yang, H.; Han, X.; Fu, A.; Li, Q.; Li, N.; Gong, S.; et al. Long-read sequencing and de novo assembly of a Chinese genome. *Nat. Commun.* **2016**, *7*, 12065. [CrossRef] [PubMed]
15. Seo, J.S.; Rhie, A.; Kim, J.; Lee, S.; Sohn, M.H.; Kim, C.U.; Hastie, A.; Cao, H.; Yun, J.Y.; Kim, J.; et al. De novo assembly and phasing of a Korean human genome. *Nature* **2016**, *538*, 243–247. [CrossRef] [PubMed]
16. Pendleton, M.; Sebra, R.; Pang, A.W.; Ummat, A.; Franzen, O.; Rausch, T.; Stutz, A.M.; Stedman, W.; Anantharaman, T.; Hastie, A.; et al. Assembly and diploid architecture of an individual human genome via single-molecule technologies. *Nat. Methods* **2015**, *12*, 780–786. [CrossRef] [PubMed]
17. Mostovoy, Y.; Levy-Sakin, M.; Lam, J.; Lam, E.T.; Hastie, A.R.; Marks, P.; Lee, J.; Chu, C.; Lin, C.; Dzakula, Z.; et al. A hybrid approach for de novo human genome sequence assembly and phasing. *Nat. Methods* **2016**, *13*, 587–590. [CrossRef] [PubMed]
18. Wong, K.H.; Levy-Sakin, M.; Kwok, P.Y. De novo human genome assemblies reveal spectrum of alternative haplotypes in diverse populations. *Nat. Commun.* **2018**, *9*, 3040. [CrossRef] [PubMed]
19. Chin, C.S.; Alexander, D.H.; Marks, P.; Klammer, A.A.; Drake, J.; Heiner, C.; Clum, A.; Copeland, A.; Huddleston, J.; Eichler, E.E.; et al. Nonhybrid, finished microbial genome assemblies from long-read SMRT sequencing data. *Nat. Methods* **2013**, *10*, 563–569. [CrossRef] [PubMed]
20. Zheng-Bradley, X.; Streeter, I.; Fairley, S.; Richardson, D.; Clarke, L.; Flicek, P. Alignment of 1000 Genomes Project reads to reference assembly GRCh38. *Gigascience* **2017**, *6*, 1–8. [CrossRef] [PubMed]
21. Kurtz, S.; Phillippy, A.; Delcher, A.L.; Smoot, M.; Shumway, M.; Antonescu, C.; Salzberg, S.L. Versatile and open software for comparing large genomes. *Genome Biol.* **2004**, *5*, R12. [CrossRef] [PubMed]
22. Sedlazeck, F.J.; Rescheneder, P.; Smolka, M.; Fang, H.; Nattestad, M.; von Haeseler, A.; Schatz, M.C. Accurate detection of complex structural variations using single-molecule sequencing. *Nat. Methods* **2018**, *15*, 461–468. [CrossRef] [PubMed]
23. Camacho, C.; Coulouris, G.; Avagyan, V.; Ma, N.; Papadopoulos, J.; Bealer, K.; Madden, T.L. BLAST+: Architecture and applications. *BMC Bioinform.* **2009**, *10*, 421. [CrossRef] [PubMed]
24. Wang, K.; Li, M.; Hakonarson, H. Annovar: Functional annotation of genetic variants from high-throughput sequencing data. *Nucleic Acids Res.* **2010**, *38*, e164. [CrossRef] [PubMed]
25. Sherry, S.T.; Ward, M.H.; Kholodov, M.; Baker, J.; Phan, L.; Smigielski, E.M.; Sirotkin, K. dbSNP: The NCBI database of genetic variation. *Nucleic Acids Res.* **2001**, *29*, 308–311. [CrossRef] [PubMed]

26. O'Leary, N.A.; Wright, M.W.; Brister, J.R.; Ciufo, S.; Haddad, D.; McVeigh, R.; Rajput, B.; Robbertse, B.; Smith-White, B.; Ako-Adjei, D.; et al. Reference sequence (RefSeq) database at NCBI: Current status, taxonomic expansion, and functional annotation. *Nucleic Acids Res.* **2015**, *44*, D733–D745. [CrossRef] [PubMed]

27. Genomes Project, C.; Auton, A.; Brooks, L.D.; Durbin, R.M.; Garrison, E.P.; Kang, H.M.; Korbel, J.O.; Marchini, J.L.; McCarthy, S.; McVean, G.A.; et al. A global reference for human genetic variation. *Nature* **2015**, *526*, 68–74.

28. Chin, C.S.; Peluso, P.; Sedlazeck, F.J.; Nattestad, M.; Concepcion, G.T.; Clum, A.; Dunn, C.; O'Malley, R.; Figueroa-Balderas, R.; Morales-Cruz, A.; et al. Phased diploid genome assembly with single-molecule real-time sequencing. *Nat. Methods* **2016**, *13*, 1050–1054. [CrossRef] [PubMed]

29. Altschul, S.F.; Gish, W.; Miller, W.; Myers, E.W.; Lipman, D.J. Basic local alignment search tool. *J. Mol. Biol.* **1990**, *215*, 403–410. [CrossRef]

30. Bennett, H.M.; Mok, H.P.; Gkrania-Klotsas, E.; Tsai, I.J.; Stanley, E.J.; Antoun, N.M.; Coghlan, A.; Harsha, B.; Traini, A.; Ribeiro, D.M.; et al. The genome of the sparganosis tapeworm *Spirometra erinaceieuropaei* isolated from the biopsy of a migrating brain lesion. *Genome Biol.* **2014**, *15*, 510. [CrossRef] [PubMed]

31. Thorvaldsdottir, H.; Robinson, J.T.; Mesirov, J.P. Integrative Genomics Viewer (IGV): High-performance genomics data visualization and exploration. *Brief. Bioinform.* **2013**, *14*, 178–192. [CrossRef] [PubMed]

32. Bickhart, D.M.; Rosen, B.D.; Koren, S.; Sayre, B.L.; Hastie, A.R.; Chan, S.; Lee, J.; Lam, E.T.; Liachko, I.; Sullivan, S.T.; et al. Single-molecule sequencing and chromatin conformation capture enable de novo reference assembly of the domestic goat genome. *Nat. Genet.* **2017**, *49*, 643–650. [CrossRef] [PubMed]

33. Jain, M.; Koren, S.; Miga, K.H.; Quick, J.; Rand, A.C.; Sasani, T.A.; Tyson, J.R.; Beggs, A.D.; Dilthey, A.T.; Fiddes, I.T.; et al. Nanopore sequencing and assembly of a human genome with ultra-long reads. *Nat. Biotechnol.* **2018**, *36*, 338. [CrossRef] [PubMed]

34. Redon, R.; Ishikawa, S.; Fitch, K.R.; Feuk, L.; Perry, G.H.; Andrews, T.D.; Fiegler, H.; Shapero, M.H.; Carson, A.R.; Chen, W.; et al. Global variation in copy number in the human genome. *Nature* **2006**, *444*, 444–454. [CrossRef] [PubMed]

35. Yuan, S.; Johnston, H.R.; Zhang, G.; Li, Y.; Hu, Y.J.; Qin, Z.S. One Size Doesn't Fit All—RefEditor: Building Personalized Diploid Reference Genome to Improve Read Mapping and Genotype Calling in Next Generation Sequencing Studies. *PLoS Comput. Biol.* **2015**, *11*, e1004448. [CrossRef] [PubMed]

36. Lander, E.S.; Linton, L.M.; Birren, B.; Nusbaum, C.; Zody, M.C.; Baldwin, J.; Devon, K.; Dewar, K.; Doyle, M.; FitzHugh, W.; et al. Initial sequencing and analysis of the human genome. *Nature* **2001**, *409*, 860–921. [PubMed]

![genes logo] **genes**

MDPI

Article

A Statistical Method for Observing Personal Diploid Methylomes and Transcriptomes with Single-Molecule Real-Time Sequencing

Yuta Suzuki [1,†], Yunhao Wang [2,†], Kin Fai Au [2,3,*] and Shinichi Morishita [1,*]

[1]   Department of Computational Biology and Medical Sciences, Graduate School of Frontier Sciences,
      The University of Tokyo, Tokyo 277-8561, Japan; yuta_suzuki@edu.k.u-tokyo.ac.jp
[2]   Department of Internal Medicine, University of Iowa, Iowa City, IA 52242, USA; yunhaowang@126.com
[3]   Department of Biomedical Informatics, Ohio State University, Columbus, OH 43210, USA
[*]   Correspondence: kinfai.au@osumc.edu (K.F.A.); moris@edu.k.u-tokyo.ac.jp (S.M.);
      Tel.: +81-47-136-3985 (S.M.)
[†]   These authors contributed equally to this work.

Received: 15 August 2018; Accepted: 12 September 2018; Published: 19 September 2018

Abstract: We address the problem of observing personal diploid methylomes, CpG methylome pairs of homologous chromosomes that are distinguishable with respect to phased heterozygous variants (PHVs), which is challenging due to scarcity of PHVs in personal genomes. Single molecule real-time (SMRT) sequencing is promising as it outputs long reads with CpG methylation information, but a serious concern is whether reliable PHVs are available in erroneous SMRT reads with an error rate of $\sim$15%. To overcome the issue, we propose a statistical model that reduces the error rate of phasing CpG site to 1%, thereby calling CpG hypomethylation in each haplotype with >90% precision and sensitivity. Using our statistical model, we examined *GNAS* complex locus known for a combination of maternally, paternally, or biallelically expressed isoforms, and observed allele-specific methylation pattern almost perfectly reflecting their respective allele-specific expression status, demonstrating the merit of elucidating comprehensive personal diploid methylomes and transcriptomes.

Keywords: statistical methods; DNA methylation; gene expression; single molecule real-time sequencing; allele-specific analysis

1. Introduction

DNA methylation plays important regulatory roles in a wide range of biological processes, including differentiation, transposon repression, and cancer progression [1–3]. Several technological advances now enable us to evaluate genome-wide DNA methylation [4] at the resolution of a single base-pair [5]. Furthermore, single-cell biology can now be applied to epigenetics, allowing methylation to be measured at the single-cell level. This creates a unique research frontier [6–8]. Despite such advances in methodology, the detection of allele-specific methylation (ASM) [9], in which only one of two homologous chromosomes is methylated in a specific region, remains challenging.

To distinguish two homologous chromosomes directly, several studies have explicitly utilized heterozygous variants, as such variants define the differences between two homologous chromosomes. One approach involves a two-step experiment [10,11]. In the first step, DNA fragments containing methylated alleles were enriched using a methylation-sensitive restriction enzyme or by methylated DNA immunoprecipitation. In the second step, sequence variants in the library were quantified using a single nucleotide polymorphism (SNP) array or DNA sequencing. Variants associated with methylation might thus be over-represented when compared with an appropriate negative control. This approach is relatively cost-effective and comprehensive, but the resolution is limited by the distribution of the relevant restriction enzyme cleavage sites, which are far sparser than CpG sites.

Another approach exploits heterozygous variants within bisulfite-treated sequencing reads [12,13]. To assign a read to one of two alleles, the read must contain at least one informative (i.e., heterozygous) variant in addition to the CpG site. However, we will show below that this condition is rarely satisfied when short bisulfite-treated reads are used; bisulfite cleaves DNA into fragments less than 500 bp long [14], with maximum read lengths of 1500 bp [15]. For example, Kuleshov et al. performed haplotyping of a genome using a read cloud containing long-range information and performed short-read bisulfite sequencing to survey ASM in a genome-wide manner; however, ASM was only partially observed [16].

It is difficult to observe comprehensive ASM for a given individual genome because of the lack of sufficient heterozygous variants available in short reads around the CpG sites. Consequently, the current genome-wide overview of ASM is an average mixture of observations for many individuals in a population.

In the present work, we hypothesize that long reads are necessary to directly observe genome-wide ASM of most CpG sites in an individual genome. We developed an alternative method that allows evaluation of regions of intermediate methylation status. We used kinetic information obtained by PacBio (Pacific Biosciences, Menlo Park, California, USA) sequencing to call regional CpG methylations as reported previously [17]. We term the allele-specific methylome data obtained using phased long reads as the personal diploid methylome, as these data are comprehensive genome-wide ASM data obtained from a single individual based on personal haplotype information.

Previous studies have revealed the prevalence of allele-specific expression (ASE) in humans and demonstrated a link between ASM and ASE [18,19]. We incorporated transcriptome data obtained using long reads (as in "Iso-seq" studies using PacBio long reads [20]) and short reads to confirm that some of the ASM statuses we detected are consistent with their transcriptional activity, including their ASE statuses, i.e., personal diploid transcriptomes.

2. Materials and Methods

2.1. Data Source

DNA sequencing data and phased variant information for HG002 were obtained from a file transfer protocol (FTP) repository of the GIAB (Genome in a Bottle) consortium [21]. The original cell lines are available (Coriell GM24385). DNA/RNA sequencing data and assembled haplotigs (haplotype A and B) for AK1 were obtained from a public repository (Accession No. PRJNA298944) [22]. Both samples were lymphoblastoid cell lines.

2.2. Assignment of Reads to Each Haplotype for AK1 and HG002

For AK1 data, contigs of haplotypes A and B were separately aligned to human hg38 reference genome by BWA aligner (version 0.7.15) [23] with the "mem" mode. The phased heterozygous single nucleotide variants (SNVs) were called by comparing the haplotype A and haplotype B genomes. For both samples, we mapped the genomic reads to the reference using BLASR (http://bix.ucsd.edu/projects/blasr/) with default options set in SMRT Analysis 2.3.0 (Pacific Biosciences) (Figure 1a). Then, if the read contained matches to phased heterozygous variants (PHVs), we counted the number of PHVs supporting each allele. Assignment of the allele for each read was determined by majority voting, and reads were excluded from further analysis if they contained none of the PHVs, or if the voting was tied. In total, 24,181,074 reads (210,782 Mb) from the AK1 dataset were aligned to hg38. The average length of the mapped reads was 8717 bp. Of the reads, 13,857,752 (139,467 Mb) contained at least one match to a PHV, and a haplotype label was assigned. Although these reads constituted 57.3% of all mapped reads in terms of read number, they contained 66.2% of the mapped bases. More bases were retained because longer reads were more likely to contain matches to the PHV. In other words, reads with no matches were likely to be shorter, therefore affecting a relatively small number of bases. Consequently, the average length of reads assigned to a haplotype was 10,064 bp, 115% that of the average length in the original dataset.

For the HG002 dataset, the phased variants calculated using the linked-read technology were available [24]. Starting from 23,031,407 reads (168,051 Mb) aligned to hg38, 13,676,974 reads (111,543 Mb) were assigned a haplotype label. Thus, we retained 59.4% of all mapped reads and 66.4% of all mapped bases. The average read length of reads assigned to a haplotype was 8156 bp, 112% that of the average length of the original dataset, which was 7296 bp.

2.3. Generating the Diploid Methylomes

We examined the single-molecule real-time (SMRT) read sets of both alleles separately and called the regional methylation status of genome-wide CpG sites using the kinetic information inherent in reads, as described previously [17]. To determine the cause of read assignment errors in the ASM detection pipeline and how they affect the accuracy of final ASM calls, we assumed that Inter-pulse duration (*IPD*) ratio statistics around the SNV are perturbed by

$$\Delta IPD = random(-1,1) \times P \times (IPD - 1.0), \tag{1}$$

where $random(-1,1)$ is sampled from the uniform distribution over $[-1,1]$ and $P = 1\%$. The third factor, $(IPD - 1.0)$, captures a typical scale of IPD deviation. The second factor (P) represents probability of read assignment error, which was estimated to be 1% (Section 3).Then, the perturbation will be largest (fully realized) when two alleles are in completely contrastive methylation state (100%/0% of the molecules methylated), which corresponds to the extreme case where the first factor $= 1$ or -1. In general, we cannot know underlying methylation states in each allele, and we must utilize random distribution to mimic the magnitude of perturbation realized. Moreover, we cannot assume a specific distribution for *IPD* perturbation, thus we decided to use uniform distribution indifferently.

2.4. Calculating the Distribution of Phased Heterozygous Variants with Respect to CpG Islands or Exons

CpG island (CGI) annotation was retrieved from the UCSC Genome Browser (https://genome. ucsc.edu/). For both PHVs and common SNPs, the distance from the CGI was calculated as the genomic distance from the center of the CGI. To calculate the distribution of PHVs and common SNPs with respect to exons, a gene model (GENCODE ver. 24) was intersected with each feature.

2.5. Allele-Specific Expression Analysis of AK1

Short-read (Illumina, San Diego, California, US) RNA-seq data was aligned to hg38 reference genome by Hisat2 aligner (version 2.0.0-beta) [25] with the default parameter. For long-read (PacBio) Iso-seq data, CCS reads were extracted from raw h5 files by SMRT Analysis software (version 2.3.0) with the parameter "–minFullPasses 0 –minPredictedAccuracy 70"; then, CCS reads were aligned to hg38 reference genome by GMAP aligner (version 2016-08-16) [26] with the default parameter. IDP-ASE software [19] was used to perform ASE analysis at the gene and isoform levels by short-read RNA-seq data and long-read Iso-Seq data.

2.6. Identifying CpG Islands with Allele-Specific Methylation

Of the 26,866 CGIs in the entire genome, we examined 20,140 with at least 30 CpGs to focus on the more functional CGIs. Of these, in the HG002 dataset, 5063 were not covered by long reads after allelic origin assignment, partly because they are relatively distant (4016 are separated by ≥5000 bp) from their nearest PHV. We required that all CGIs be covered by a sufficient number (≥16.0× for each haploid) of long reads to reduce the false discovery rate [17]. A total of 7093 CGIs met these criteria. Similarly, we analyzed the AK1 dataset and determined that 7322 (of 20,140) CGIs were not covered by any read, but 10,087 of the remaining 12,818 CGIs had sufficient coverage (≥16.0×).

We then examined the methylation status of each CpG site using our methylation detection algorithm AgIn [17]. Specifically, we calculated the methylation level of the CGI as the (unweighted) average of the methylation status, presented as 0 or 1 (unmethylated or methylated, respectively), within the CGI.After identifying ASM CGIs, each CGI was associated with a gene(s) by manual inspection on a genome browser, and the closest gene (transcript) was recorded.

3. Results

3.1. A Statistical Model for Accurate Read Assignment

The first step towards ASM detection using long reads is read assignment (Figure 1a), i.e., to assign each read to an original allele based on variants contained in the read. This is a nontrivial task considering that long reads have a higher error rate. While the consensus accuracy of long reads can be sufficiently good for detecting small genomic variants, here we are faced with a raw read error rate as we assign each single read to an allele one by one. It is difficult to call heterozygous variants (found in ~0.1% of all genomic positions) using only long reads with an error rate of, for example, ~15%.

We instead utilized highly accurate haplotype PHVs determined by short read sequencing to achieve reliable read assignment. By this design, we were able to ignore discordance between the reference genome and long reads on non-variant sites, which effectively corrected most of the errors in the long reads. Another technique employed to improve read assignment was to use only SNVs and to ignore insertions and deletions (indels), which constitute a dominant fraction of the errors in PacBio long reads, thereby reducing the relevant error rate. This observation has been utilized by several authors to handle erroneous long reads, in diploid-aware de novo assembly [27] and in hybrid error-correction procedure [28], but no explicit modelling has been done for the error probability of long reads phasing. As we will see, even after omitting indels from the analysis, sufficient SNVs are available for read assignment to haplotypes. PacBio read assignment for transcripts (Iso-seq reads) can be similarly performed using heterozygous variants found in exons (Figure 1b).

We can now calculate the expected error rate of the read assignment. Suppose there is only one heterozygous SNV within a read. According to the typical error profile, here we assumed we would observe erroneous insertion (10%), deletion (5%), or substitution to any of the wrong bases (3%) at each position in the PacBio read [29].

Then, read assignment error occurs if and only if the base at an SNV site is substituted by one of the other three bases that incidentally supports the other allele with a probability of 1% (3% substitution rate divided by three). Thus, the error rate of read assignment is expected to be approximately 1% for a read with one SNV. Other types of sequencing errors, indels, and substitutions to bases other than one supporting the wrong allele do not cause read assignment error because such reads are not assigned to either allele.

With more than one heterozygous SNV within a read, the accuracy of read assignment improves drastically. If there are two heterozygous SNVs, then there are two cases where read assignment should occur: First, simultaneous substitution errors resulting in the bases of the wrong allele occur at both sites, giving two variants supporting the wrong allele. The probability of such an event is $1\% \times 1\% = 1.0 \times 10^{-4}$; Second, one of the SNVs is lost by deletion (5%) or another substitution (2%) in the read, and specific substitution occurs at the other site. In this case, only one variant, which supports the wrong allele, is observed in the read. The probability for this event is $\binom{2}{1} \times (5\% + 2\%) \times 1\% = 1.4 \times 10^{-3}$. As these two cases are mutually exclusive, the read assignment error rate is $1.0 \times 10^{-4} + 1.4 \times 10^{-3} = 1.5 \times 10^{-3} = 0.15\%$ for a two-SNV read.

Generally, read assignment error occurs when the number of SNVs supporting the wrong allele is greater than that of SNVs supporting the correct allele. If the read has N SNVs, and k SNVs are missed, then the read would be assigned to the wrong allele if more than half (denoted by l below) of the remaining $(N - k)$ SNVs support the wrong allele. Thus, the probability of assignment error for a read with N SNVs is

$$\sum_{k=0}^{N-1} \binom{N}{k} (5\% + 2\%)^k \{ \sum_{l=\lceil (N-k)/2 \rceil}^{N-k} \binom{N - k}{l} (1\%)^l \}, \tag{2}$$

which is approximately $0.15\%, 0.047\%$, and 0.010%, when $N = 2, 3$, and 4, respectively. Therefore, the overall assignment error rate decreases exponentially with the number of available heterozygous SNVs (Figure 1c).

Due to the limited availability of a ground truth dataset for personal diploid methylomes, it is difficult to quantitatively assess the accuracy of ASM detection using experimental data. Therefore, we approximated the accuracy by considering major potential sources of errors. To examine the effect of read assignment errors, we added random perturbation, which was proportional to the frequency of read assignment errors, to the IPD ratio of every position independently. Using this setting, the predictive performance of the method was almost unchanged (both the sensitivity and precision were >90%; Figure 1d), presumably because random errors were averaged out in our read assignment. While our analysis simplifies the real situation, it conveys why sequencing errors would not severely affect the accuracy of the method contrary to the impression. Therefore, we conclude that the major source of inaccuracy of the method is methylation detection itself, which is guaranteed to be highly accurate (>93%) [17].

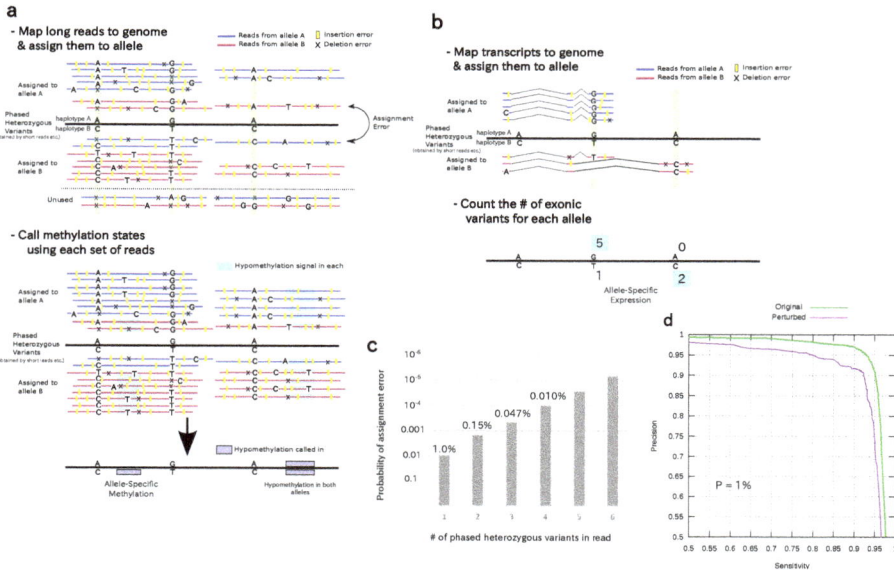

Figure 1. *Cont.*

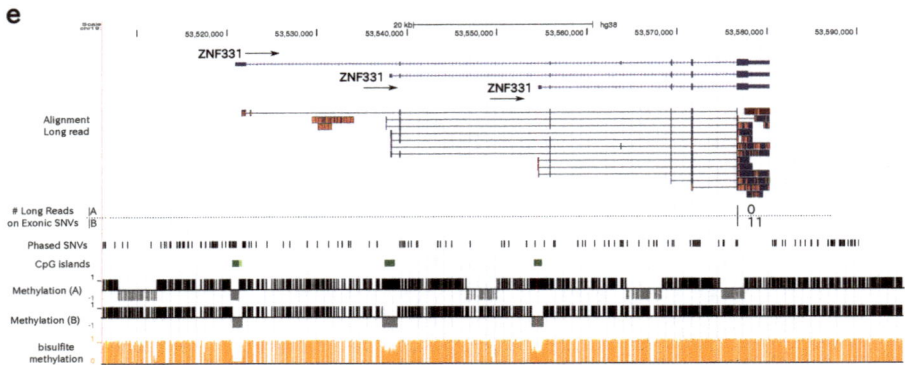

Figure 1. Outline of the proposed method to detect allele-specific methylation (ASM). (**a**) In our method, we assume that haplotype information for the genome is available, i.e., phased heterozygous variants (PHVs) exist (horizontal line in the middle with letters indicating PHVs), which serve as sites of interest (sites shaded in green) for the allele assignment process. Available heterozygous variants found in reads are indicated by letters (A, C, G, or T) at the shaded sites. Other mismatches and insertions/deletions (indels) (shown as letters and yellow blocks) in reads can be assumed to be sequencing errors and therefore ignored. After assigning PacBio reads to either allele (haplotype) by PHVs on the reads, the methylation status of each allele is predicted using (average) kinetics data obtained in the PacBio sequencing process (shaded in blue). Of note, if ASM is not present in the region, wrong assignment of reads does not affect the accuracy of the methylation call; (**b**) Outline of the detection of allele-specific expression (ASE). Only exonic PHVs (two of three in this figure) can be used to distinguish two alleles. Next, ASE can be detected as an imbalance of alleles observed in reads. (**c**) The probability of allele assignment error depends on the number of available PHVs in the reads (*x*-axis). The probability was calculated using an equation presented in the text and is shown in logarithmic scale on the *y*-axis; (**d**) prediction performance (sensitivity and precision) of the method for assessing the perturbed inter-pulse duration (IPD) ratio (purple line). For comparison, typical performance statistics for the original IPD are shown (green line). "$P = 1\%$" indicates that the IPD was perturbed to simulate 1% read assignment error, as described in the text; (**e**) Example of a region exhibiting ASM in diploid methylomes and transcriptomes. On the right: two CpG islands (CGIs) are shown in the middle; one allele (labeled A) is methylated and the other (B) unmethylated. Each CGI overlaps with the promoter regions of distinct isoforms of a known imprinted gene *ZNF331*. Bisulfite sequencing data in the bottom track exhibited intermediate-level methylation for the two CGIs showing ASM. From top to bottom, the panel shows the following features: gene structure, alignments of long RNA-seq (Iso-seq) reads, RNA-seq read counts for two alleles, which indicates ASE, sites of PHVs available in this personal genome (black marks), which were used to determine the allelic origins of the sequencing reads, annotated CGIs (green rectangles), methylation levels of the CpG sites of two alleles that were predicted using single-molecule real-time (SMRT) reads (respective black and gray bars towards positive and negative indicate methylated and unmethylated, respectively), and publicly available data on methylation levels via bisulfite sequencing (orange bars).

3.2. Generating Diploid Methylomes and Transcriptomes for the AK1 and HG002 Datasets

To demonstrate our method of calling ASM, we used two independent datasets: AK1 (Asian Korean) [22] and HG002 (Ashkenazi Trio son) [30], according to the procedure illustrated in Figure 1a (see the details in Methods). The resulting set of methylation calls is a personal diploid methylome, as it comprises two methylomes, each representing one haploid. For the AK1 dataset, RNA-seq data obtained using long and short reads were available and thus used to support the hypothesis that the differential methylation between two alleles that we detected was associated with differential transcription activity. RNA-seq data were mapped to the genome, and then the number of reads supporting transcription from each allele was recorded (Figure 1b), building a pair of transcriptomes in two homologous chromosomes, which we call personal diploid transcriptomes.

Figure 1e shows an example of ASM detected using our method in the genomic region encoding an imprinted gene, *ZNF331*. There are three CGIs in this region, and each CGI corresponds to a promoter region in distinct isoforms of the *ZNF331* gene. While the CGI to the left in the panel was unmethylated for both alleles, the other two CGIs (in the middle and to the right) showed ASM, and our methylation calls informed us that the same allele (i.e., allele B) was unmethylated. Of note, a publicly available methylation-level annotation[31] from a different sample by bisulfite sequencing suggested that these two CGIs are in an intermediate methylation status. The alignment of long-read transcripts and read counts at the exonic PHV supported that the corresponding two isoforms were transcribed exclusively from allele B. Thus, these results suggest that the detected ASM is correlated with the transcriptional activity of the genes. We will cover other examples in later sections to generalize this observation.

3.3. Distribution of Phased Heterozygous Variants in Two Personal Genomes

We next examined how the possibility of assigning reads and CpGs into alleles would be limited by the distribution of PHVs in personal genomes. Specifically, given a read of length *l* bp containing a CpG site, the allelic origin of the read can be determined only when the nearest PHV is located within *l* bp from that site, and both the CpG site and PHV are covered by the same single read. Therefore, to assess this approach, we calculated the proportions of CpGs or CGIs located within a specific distance from the nearest PHV (Figure 2a,b). Of note, these factors depend on the distribution of PHVs available in a given sample and thus can be quite different among individual samples.

One may use the set of SNPs for which the minor allele frequencies are $\geq$5% in at least 1 of 26 major populations of dbSNPs [32]. As we observed that 83.0% of CpG sites are located within 500 bp from common SNPs (Figure 2a), it would be possible for relatively short reads to determine the allelic origin of themselves if these common SNPs are heterozygously present in an individual genome. However, the conditions posed by the real distribution of PHVs are more severe. Indeed, in the AK1 (HG002) dataset, at most 11.3% (12.3%), 33.7% (37.5%) and 46.1% (51.1%) of CpGs were apparent using read lengths of 100, 500 and 1000 bp, respectively, whereas 72.2% (81.3%) of CpGs were apparent using a read length of 8000 bp (Figure 2a,b, purple line). Therefore, longer reads are essential to detect ASM in real-world situations.

When we try to detect ASE using these PHVs, the only variants we can rely on are those appearing in the RNA-seq reads, i.e., exonic variants. Thus, the situation becomes even more difficult for ASE analysis. For the AK1 dataset, 46.2% of the PHVs were found in intergenic regions and 50.0% in intronic regions (Figure 2c). Thus, only the remaining 3.8% of variants were present in exons to call ASE in this individual. Similarly, 89.1% of ~310 k exons do not contain such variants, limiting the possibility of determining the expressing allele. While the low abundance of PHVs within exons is largely explained by the fact that exons constitute only a small fraction of the genome, the density of PHVs is also smaller in exons, presumably because of selective pressure on the coding sequences. On average, there were 0.68 PHVs within exons (0.79 PHVs within introns) per 1 kbp.

3.4. Allele-Specific Methylation of CpG Islands and Allele-Specific Expression

Given the fact that exonic phased variants are found only in a small number of transcripts, genome-wide observation of ASM would provide alternative information about the transcriptional activity of individual genomes. To prove this concept, we generated diploid methylomes for the AK1 dataset and compared it with ASE analysis of RNA-seq data for the same dataset (Figure 2d,e).

Shown in the middle of the first panel, a CGI is located in the promoter region of *ZNF597* (Figure 2d). We detected ASM around this CGI, and allele A was unmethylated upstream of the transcription start site (TSS) of *ZNF597*; thus, we predicted that the gene is expressed exclusively from allele A. Consistent with the prediction, we found two long reads and 59 short reads, supporting transcription from allele A, while no reads supported the other allele, B. This identification of ASE was based on an exonic SNV within the last exon, which was the only exonic SNV present in this region, highlighting the sparseness of PHVs in exons.

The second example region is the *GNAS* complex locus, where several isoforms of the *GNAS* gene show ASE (Figure 2e) [33]. Specifically, while *Gsα* (shown at the right end) is expressed from both alleles, the *A/B* transcripts, *XLαs* is paternally expressed (allele B in Figure 2e), and *NESP55* is maternally expressed (allele A). We confirmed that *Gsα* was expressed from both alleles, as the exon specific to the isoform contained a PHV, and both alternative alleles were observed in the RNA-seq reads. For the other isoforms, although we could not directly observe their allelic expression patterns due to the lack of phased variants in exonic regions, the CGIs located in the promoter regions of each isoform showed an ASM pattern consistent with the expected expression pattern; the two CGI regions in the promoters of the *A/B* transcripts, *XLαs*, and *GNAS-AS* were allele-specifically methylated on the same allele, *B*, and the CGI at the promoter of *NESP55* was methylated on the other allele, *A*. Thus, we could predict the expected expression pattern for this locus via its methylation pattern.

These examples demonstrate that the detected ASM status of CGIs can reflect the expression status of the corresponding genes/isoforms. Therefore, such ASM of CGIs would provide useful information, especially when ASE is difficult to detect because of the absence of phased variant sites within exons.

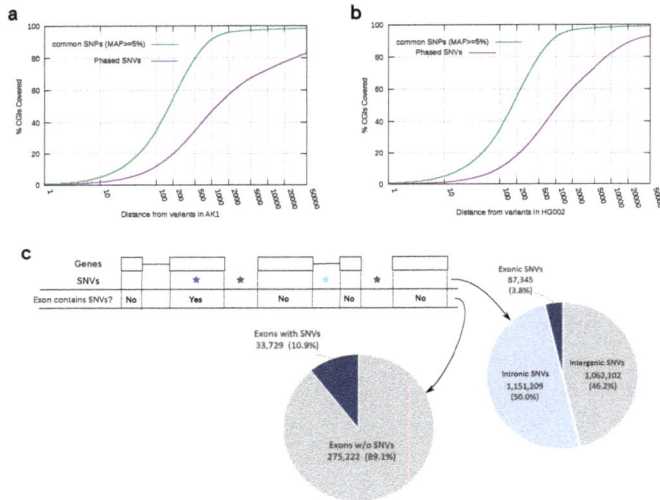

Figure 2. *Cont.*

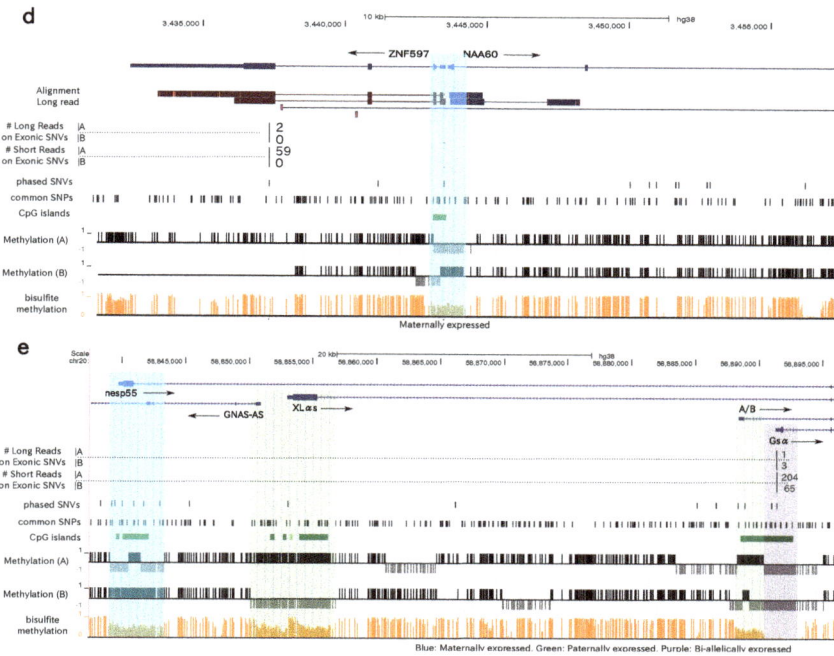

Figure 2. (**a**,**b**) Proportions of CGIs located within a distance in the *x*-axis from the nearest genomic feature, common single nucleotide polymorphisms (SNPs) (green) and heterozygous single nucleotide variants (hetSNVs, or PHVs) (purple), in each genome of (**a**) AK1 and (**b**) HG002. Common SNPs and PHVs distributed differently in both personal genomes. PHVs were essential in determining the proportions of CGIs; (**c**) distribution of PHVs with respect to exons. The left pie chart shows the proportion of exons containing PHVs for which the ASE status can be assessed directly. The right pie chart shows the ratios of PHVs in exonic (blue), intronic (pale blue), or intergenic (gray) regions, thus classifying PHVs into three categories; (**d**) example showing personal diploid methylomes and transcriptomes in the AK1 genome. The CGI in the bidirectional promoter region (area shaded in blue) of the *ZNF597* and *NAA60* genes showed ASM. The RNA-seq reads (both long and short) support that transcription was only derived from allele A, which is the unmethylated allele in the region; (**e**) Personal diploid methylomes around the *GNAS* complex locus in the AK1 genome. The four regions are colored to show their known transcriptional pattern: maternally expressed (blue), paternally expressed (green), or expressed from both alleles (purple). Correspondingly, these regions shaded with different colors exhibited distinct methylation patterns. Of note, the ASM regions exhibited an intermediate level of methylation according to bisulfite sequencing (bottom). RNA-seq reads suggested the expression of Gsα from both alleles.

3.5. Statistics of Allele-Specific Methylation CpG Islands

Applying the same methodology to the HG002 dataset, we determined the methylation status of genome-wide CGIs by summarizing the allelic methylation status of CpG sites contained in each CGI (Methods). We calculated the methylation score for each CGI as the average of the methylation scores of all CpGs comprising the CGI (Figure 3a). Next, we selected the 70 CGIs with the top 1% absolute differences ($\geq$0.68) in methylation scores between the haploids of the diploid methylomes (Table S1). For comparison, we analyzed the AK1 dataset and determined 139 CGIs (1.3%) had a methylation difference $\geq$0.68, which is consistent with the ratio in the HG002 dataset. Thus, we continued our analysis using HG002 data. We noted that the distances between these ASM CGIs and the PHVs were not necessarily small. Of the 70 ASM CGIs, 28 were separated by $\geq$500 bp and 9 by $\geq$1000 bp from their nearest PHV, which meant that the methylation status of the alleles of these CGIs could not be measured simultaneously when the reads were shorter than 500 or 1000 bp.

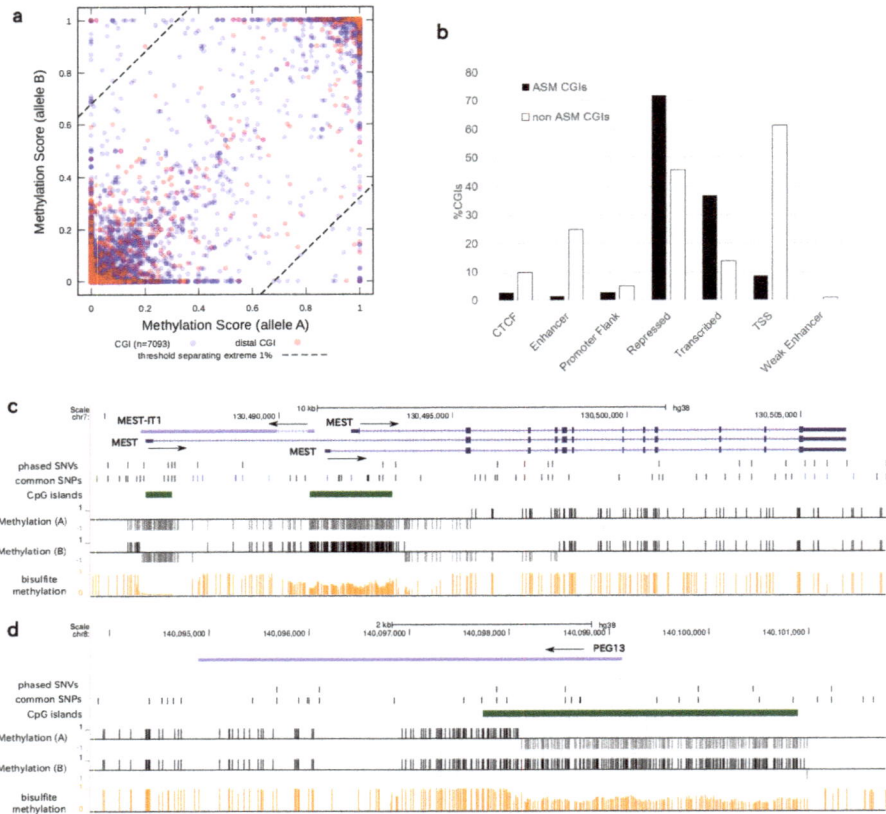

Figure 3. (**a**) Summary of the methylation scores (for each allele) for CGIs in personal diploid methylomes in HG002. Each CGI is shown as a circle. On the two opposite corners (top left and bottom right), CGIs with the top 1% absolute differences in methylation levels between the two alleles were provisionally classified as ASM CGIs. Red circles: The corresponding CGIs were separated from the nearest PHV by 1000 bp or more. Blue circles: Separation <1000 bp; (**b**) distribution of each type of CGI, ASM (black bar) or non-ASM (white bar), with respect to the functional annotation of genomic regions; (**c**) example showing personal diploid methylomes in the *MEST* gene-coding region of the HG002 (Ashkenazim Trio Son) genome. Although the upstream CGI (with 66 CpG sites) was unmethylated in both alleles, the larger downstream CGI (with 184 CpG sites) exhibited ASM. The CGIs corresponded to the promoter regions of different isoforms of the genes; (**d**) another example of ASM around an imprinted gene *PEG13*, paternally expressed gene 13.

To confirm that the detected ASM CGIs are functionally relevant, we compared the ASM CGIs with genomic annotations from the combined segmentation by Segway and ChromHMM defined in the ENCODE project (Figure 3b) [34]. Of note, CGIs in general significantly overlapped the TSS, as expected. In contrast to this background, CGIs showing ASM overlapped with segments annotated as "transcribed regions" or "repressed regions" more than with the TSS. This result may seem somewhat contradictory, since any single gene cannot be both transcribed and repressed at the same time; however, this still may be a plausibly correct categorization for regions containing ASM genes because they can be, by definition, in two contrasting states in each of the alleles.

We also confirmed that our list of candidate ASM CGIs contains a number of CGIs overlapping with known imprinted genes. For example, we reproduced the expected ASM around known imprinted genes such as *MEST* (Figure 3c), *PEG13* (Figure 3d), *HYMAI*, and *ZNF597*, etc. (Table S1) [35]. Indeed, CGIs with a larger methylation difference between two alleles were enriched with imprinted genes ($p = 0.007$, U test). Thus, we again confirmed the validity of our method by successfully recovering the imprinted genes as ASM regions.

4. Discussion

In this work, we examined personal diploid methylomes to directly characterize the ASM status of genome-wide CGIs based on a set of PHVs specific to each sample. A crucial technical problem in this study was the accurate assignment of erroneous reads to haplotypes. The simulation revealed, however, that the accuracy would not be severely affected by sequencing errors, as they are random in nature [29]. On the other hand, wrong SNV calls/phasing can be a source of biased errors, which should be alleviated by using a PHV set of better quality. Therefore, the overall accuracy of detected methylation statuses would essentially replicate the prediction performance of the original methylation detection method, e.g., >90% for regions with sufficient sequencing depth, for example, 20-fold on each allele [17].

Compared with previous studies employing short read sequencing [12,13,16], one novelty of our approach is that we called methylation using kinetic information from long SMRT reads. We did not employ any chemical treatments, such as bisulfite conversion, which cleaves DNA into small fragments of <1500 bp [15]. By this design, we fully exploited the lengths of the PacBio reads (>8000 bp in our data). We determined the allelic origins of more than half of the sequencing data. Thus, we were able to cover more CpGs in the genome. We previously reported that a read coverage of ~20-fold is required for detecting regional CpG methylation [17]. In cases in which sufficient read coverage is available for each allele after separation, this read coverage value can include a margin, since some reads will not contain any informative PHV and will be filtered out. Therefore, 40–50-fold of reads would be sufficient for the detection of ASM.

Compared with other existing methods, ours does not rely on the availability of restriction sites around the CpG sites, which makes our method potentially more comprehensive. Another important advantage of our method is that long reads enable the detection of ASM associated with distal heterozygous variants, and we demonstrated that such cases are not necessarily rare, as illustrated in Figure 2a,b.

As we demonstrated, comprehensive information about the genome-wide ASM status may complement ASE observations if we assume that the methylation status of promoter CGIs are correlated with transcriptional activity. We described a couple of examples to support this hypothesis using RNA-seq data in the AK1 dataset and indicated that the analysis of ASM could recapture some imprinted genes in the HG002 dataset. While we cannot use epigenetic observation as a complete surrogate for expression data, it can complement the ASE statuses of transcripts when they are more difficult to observe due to lack of exonic PHVs.

We also demonstrated that long reads are essential for the study of ASM, given the sparse distribution of heterozygous variants within individuals. Availability of haplotype-resolved variants data along the genomes was essential for the method presented, and the advent of linked-reads technology (10× GemCode/Chromium) enabled us to extract long-range (100 kbp) co-occurrences of DNA sequences for that purpose [24] and also for accuracy of the haplotype-phasing matters; with phasing errors, the sensitivity of our method might be undermined, or methylation statuses might be swapped between the two alleles. While we recommend to try such linked-read technology first as it is less labor-intensive and it can simply provide additional linking information to conventional short read sequencing data, several available methods for chromosome-scale haplotype reconstruction have been developed in recent years; there are novel protocols such as Strand-seq [36], and CPTv2-seq [37], and/or one can combine existing techniques such as Hi-C [38], optical mapping (Bionano, Tulsa, Oklahoma, USA) [39], and long/short read sequencing as well as linked-read [36,39]. Such technology renders it easier to sequence individual genomes in a manner that the majority of variants are haplotype-phased. The more accessible the haplotype-phased genomes become, the more reasonable it becomes to study the epigenome, being aware of the existence of two alleles.

Supplementary Materials: The following are available at http://www.mdpi.com/2073-4425/9/9/460/s1, Table S1: Detected CpG islands with allele-specific methylation and associated genes.

Author Contributions: Conceptualization, Y.S. and S.M.; Methodology, Y.S. and Y.W.; Software, Y.S. and Y.W.; Writing—Original Draft Preparation, Y.S.; Writing—Review and Editing, Y.S., Y.W., K.F.A. and S.M.; Supervision, K.F.A. and S.M.

Funding: This work was funded in part by Japan Society for the Promotion of Science (Grant-in-Aid for JSPS Fellows 15J03645) (Y.S.), Japan Science and Technology Agency (CREST JPMJCR13W3) (S.M.), Japan Agency for Medical Research and Development (GRIFIN) (S.M.), National Human Genome Research Institute (R01HG008759) (Y.W. and K.F.A.), Dept. of Internal Medicine, University of Iowa (institutional fund) (Y.W. and K.F.A.), and Pharmaceutical Research and Manufacturers of America Foundation (Research Starter Grant) (Informatics) (Y.W.)

Acknowledgments: The authors thank the Genome in a Bottle (GIAB) consortium for allowing us to use the Ashkenazi trio data. We also thank Michael Schnall-Levin of 10X Genomics for the stimulating discussions.

Conflicts of Interest: The authors declare no conflict of interest.

Abbreviations

The following abbreviations are used in this manuscript:

ASM	Allele-specific methylation
ASE	Allele-specific expresssion
SNP	Single nucleotide polymorphism
SNV	Single nucleotide variant
PHV	Phased heterozygous variant
IPD	Inter-pulse duration
CGI	CpG island

References

1. Jones, P.A. Functions of DNA methylation: Islands, start sites, gene bodies and beyond. *Nat. Rev. Genet.* **2012**, *13*, 484–492. [CrossRef] [PubMed]
2. Smith, Z.D.; Meissner, A. DNA methylation: Roles in mammalian development. *Nat. Rev. Genet.* **2013**, *14*, 204–220. [CrossRef] [PubMed]
3. Schübeler, D. Function and information content of DNA methylation. *Nature* **2015**, *517*, 321–326. [CrossRef] [PubMed]
4. Down, T.A.; Rakyan, V.K.; Turner, D.J.; Flicek, P.; Li, H.; Kulesha, E.; Graef, S.; Johnson, N.; Herrero, J.; Tomazou, E.M.; et al. A Bayesian deconvolution strategy for immunoprecipitation-based DNA methylome analysis. *Nat. Biotechnol.* **2008**, *26*, 779–785. [CrossRef] [PubMed]
5. Lister, R.; Pelizzola, M.; Dowen, R.H.; Hawkins, R.D.; Hon, G.; Tonti-Filippini, J.; Nery, J.R.; Lee, L.; Ye, Z.; Ngo, Q.M.; et al. Human DNA methylomes at base resolution show widespread epigenomic differences. *Nature* **2009**, *462*, 315–322. [CrossRef] [PubMed]
6. Smallwood, S.A.; Lee, H.J.; Angermueller, C.; Krueger, F.; Saadeh, H.; Peat, J.; Andrews, S.R.; Stegle, O.; Reik, W.; Kelsey, G. Single-cell genome-wide bisulfite sequencing for assessing epigenetic heterogeneity. *Nat. Methods* **2014**, *11*, 817–820. [CrossRef] [PubMed]
7. Gravina, S.; Dong, X.; Yu, B.; Vijg, J. Single-cell genome-wide bisulfite sequencing uncovers extensive heterogeneity in the mouse liver methylome. *Genome Biol.* **2016**, *17*, 1–8. [CrossRef] [PubMed]
8. Clark, S.J.; Lee, H.J.; Smallwood, S.A.; Kelsey, G.; Reik, W. Single-cell epigenomics: Powerful new methods for understanding gene regulation and cell identity. *Genome Biol.* **2016**, *17*, 72. [CrossRef] [PubMed]

9. Yamada, Y.; Watanabe, H.; Miura, F.; Soejima, H.; Uchiyama, M.; Iwasaka, T.; Mukai, T.; Sakaki, Y.; Ito, T. A comprehensive analysis of allelic methylation status of CpG islands on human chromosome 21q. *Genome Res.* **2004**, *14*, 247–266. [CrossRef] [PubMed]
10. Kerkel, K.; Spadola, A.; Yuan, E.; Kosek, J.; Jiang, L.; Hod, E.; Li, K.; Murty, V.V.; Schupf, N.; Vilain, E.; et al. Genomic surveys by methylation-sensitive SNP analysis identify sequence-dependent allele-specific DNA methylation. *Nat. Genet.* **2008**, *40*, 904–908. [CrossRef] [PubMed]
11. Schalkwyk, L.C.; Meaburn, E.L.; Smith, R.; Dempster, E.L.; Jeffries, A.R.; Davies, M.N.; Plomin, R.; Mill, J. Allelic skewing of DNA methylation is widespread across the genome. *Am. J. Hum. Genet.* **2010**, *86*, 196–212. [CrossRef] [PubMed]
12. Shoemaker, R.; Deng, J.; Wang, W.; Zhang, K. Allele-specific methylation is prevalent and is contributed by CpG-SNPs in the human genome. *Genome Res.* **2010**, *20*, 883–889. [CrossRef] [PubMed]
13. Gertz, J.; Varley, K.E.; Reddy, T.E.; Bowling, K.M.; Pauli, F.; Parker, S.L.; Kucera, K.S.; Willard, H.F.; Myers, R.M. Analysis of DNA methylation in a three-generation family reveals widespread genetic influence on epigenetic regulation. *PLoS Genet.* **2011**, *7*, e1002228. [CrossRef] [PubMed]
14. Miura, F.; Enomoto, Y.; Dairiki, R.; Ito, T. Amplification-free whole-genome bisulfite sequencing by post-bisulfite adaptor tagging. *Nucl. Acids Res.* **2012**, *40*, e136. [CrossRef] [PubMed]
15. Yang, Y.; Sebra, R.; Pullman, B.S.; Qiao, W.; Peter, I.; Desnick, R.J.; Geyer, C.R.; DeCoteau, J.F.; Scott, S.A. Quantitative and multiplexed DNA methylation analysis using long-read single-molecule real-time bisulfite sequencing (SMRT-BS). *BMC Genom.* **2015**, *16*, 350, doi:10.1186/s12864-015-1572-7. [CrossRef] [PubMed]
16. Kuleshov, V.; Xie, D.; Chen, R.; Pushkarev, D.; Ma, Z.; Blauwkamp, T.; Kertesz, M.; Snyder, M. Whole-genome haplotyping using long reads and statistical methods. *Nat. Biotechnol.* **2014**, *32*, 261. [CrossRef] [PubMed]
17. Suzuki, Y.; Korlach, J.; Turner, S.W.; Tsukahara, T.; Taniguchi, J.; Qu, W.; Ichikawa, K.; Yoshimura, J.; Yurino, H.; Takahashi, Y.; et al. AgIn: Measuring the landscape of CpG methylation of individual repetitive elements. *Bioinformatics* **2016**, *32*, 2911–2919. [CrossRef] [PubMed]
18. Schultz, M.D.; He, Y.; Whitaker, J.W.; Hariharan, M.; Mukamel, E.A.; Leung, D.; Rajagopal, N.; Nery, J.R.; Urich, M.A.; Chen, H.; et al. Human body epigenome maps reveal noncanonical DNA methylation variation. *Nature* **2015**, *523*, 212. [CrossRef] [PubMed]
19. Deonovic, B.; Wang, Y.; Weirather, J.; Wang, X.J.; Au, K.F. IDP-ASE: Haplotyping and quantifying allele-specific expression at the gene and gene isoform level by hybrid sequencing. *Nucl. Acids Res.* **2017**, *45*, e32. [CrossRef] [PubMed]
20. Au, K.F.; Sebastiano, V.; Afshar, P.T.; Durruthy, J.D.; Lee, L.; Williams, B.A.; van Bakel, H.; Schadt, E.E.; Reijo-Pera, R.A.; Underwood, J.G.; et al. Characterization of the human ESC transcriptome by hybrid sequencing. *Proc. Natl. Acad. Sci. USA* **2013**, *110*, E4821–E4830. [CrossRef] [PubMed]
21. Genome in a Bottle Consortium. Available online: ftp://ftp-trace.ncbi.nlm.nih.gov/giab/ftp/release (accessed on 18 September 2018).
22. Seo, J.S.; Rhie, A.; Kim, J.; Lee, S.; Sohn, M.H.; Kim, C.U.; Hastie, A.; Cao, H.; Yun, J.Y.; Kim, J.; et al. De novo assembly and phasing of a Korean human genome. *Nature* **2016**, *538*, 243–247. [CrossRef] [PubMed]
23. Li, H.; Durbin, R. Fast and accurate long-read alignment with Burrows–Wheeler transform. *Bioinformatics* **2010**, *26*, 589–595. [CrossRef] [PubMed]
24. Zheng, G.X.; Lau, B.T.; Schnall-Levin, M.; Jarosz, M.; Bell, J.M.; Hindson, C.M.; Kyriazopoulou-Panagiotopoulou, S.; Masquelier, D.A.; Merrill, L.; Terry, J.M.; et al. Haplotyping germline and cancer genomes with high-throughput linked-read sequencing. *Nat. Biotechnol.* **2016**, *34*, 303–311. [CrossRef] [PubMed]
25. Kim, D.; Langmead, B.; Salzberg, S.L. HISAT: A fast spliced aligner with low memory requirements. *Nat. Methods* **2015**, *12*, 357. [CrossRef] [PubMed]
26. Wu, T.D.; Watanabe, C.K. GMAP: A genomic mapping and alignment program for mRNA and EST sequences. *Bioinformatics* **2005**, *21*, 1859–1875. [CrossRef] [PubMed]
27. Chin, C.S.; Peluso, P.; Sedlazeck, F.J.; Nattestad, M.; Concepcion, G.T.; Clum, A.; Dunn, C.; O'Malley, R.; Figueroa-Balderas, R.; Morales-Cruz, A.; et al. Phased diploid genome assembly with single-molecule real-time sequencing. *Nat. Methods* **2016**, *13*, 1050. [CrossRef] [PubMed]
28. Au, K.F.; Underwood, J.G.; Lee, L.; Wong, W.H. Improving PacBio long read accuracy by short read alignment. *PLoS ONE* **2012**, *7*, e46679. [CrossRef] [PubMed]

29. Ono, Y.; Asai, K.; Hamada, M. PBSIM: PacBio reads simulator—Toward accurate genome assembly. *Bioinformatics* **2012**, *29*, 119–121. [CrossRef] [PubMed]

30. Zook, J.; Catoe, D.; McDaniel, J.; Vang, L.; Spies, N.; Sidow, A.; Weng, Z.; Liu, Y.; Mason, C.; Alexander, N.; et al. Extensive sequencing of seven human genomes to characterize benchmark reference materials. *Sci. Data* **2016**, *3*, 160025. [CrossRef] [PubMed]

31. Stunnenberg, H.G.; Abrignani, S.; Adams, D.; de Almeida, M.; Altucci, L.; Amin, V.; Amit, I.; Antonarakis, S.E.; Aparicio, S.; Arima, T.; et al. The International Human Epigenome Consortium: A blueprint for scientific collaboration and discovery. *Cell* **2016**, *167*, 1145–1149. [CrossRef] [PubMed]

32. Sherry, S.T.; Ward, M.H.; Kholodov, M.; Baker, J.; Phan, L.; Smigielski, E.M.; Sirotkin, K. dbSNP: The NCBI database of genetic variation. *Nucl. Acids Res.* **2001**, *29*, 308–311. [CrossRef] [PubMed]

33. Bastepe, M. The GNAS locus: Quintessential complex gene encoding Gsα, XLαs, and other imprinted transcripts. *Curr. Genom.* **2007**, *8*, 398–414. [CrossRef] [PubMed]

34. Consortium, T.E.P. An integrated encyclopedia of DNA elements in the human genome. *Nature* **2012**, *489*, 57–74. [CrossRef] [PubMed]

35. Baran, Y.; Subramaniam, M.; Biton, A.; Tukiainen, T.; Tsang, E.K.; Rivas, M.A.; Pirinen, M.; Gutierrez-Arcelus, M.; Smith, K.S.; Kukurba, K.R.; et al. The landscape of genomic imprinting across diverse adult human tissues. *Genome Res.* **2015**, *25*, 927–936. [CrossRef] [PubMed]

36. Porubsky, D.; Garg, S.; Sanders, A.D.; Korbel, J.O.; Guryev, V.; Lansdorp, P.M.; Marschall, T. Dense and accurate whole-chromosome haplotyping of individual genomes. *Nat. Commun.* **2017**, *8*, 1293. [CrossRef] [PubMed]

37. Zhang, F.; Christiansen, L.; Thomas, J.; Pokholok, D.; Jackson, R.; Morrell, N.; Zhao, Y.; Wiley, M.; Welch, E.; Jaeger, E.; et al. Haplotype phasing of whole human genomes using bead-based barcode partitioning in a single tube. *Nat. Biotechnol.* **2017**, *35*, 852. [CrossRef] [PubMed]

38. Ben-Elazar, S.; Chor, B.; Yakhini, Z. Extending partial haplotypes to full genome haplotypes using chromosome conformation capture data. *Bioinformatics* **2016**, *32*, i559–i566. [CrossRef] [PubMed]

39. Mostovoy, Y.; Levy-Sakin, M.; Lam, J.; Lam, E.T.; Hastie, A.R.; Marks, P.; Lee, J.; Chu, C.; Lin, C.; Džakula, Ž.; et al. A hybrid approach for de novo human genome sequence assembly and phasing. *Nat. Methods* **2016**, *13*, 587. [CrossRef] [PubMed]

![genes logo] **genes**

MDPI

Article

Single-Molecule Real-Time (SMRT) Full-Length RNA-Sequencing Reveals Novel and Distinct mRNA Isoforms in Human Bone Marrow Cell Subpopulations

Anne Deslattes Mays [1,2], **Marcel Schmidt** [1], **Garrett Graham** [1], **Elizabeth Tseng** [3], **Primo Baybayan** [3], **Robert Sebra** [4], **Miloslav Sanda** [1], **Jean-Baptiste Mazarati** [1,5], **Anna Riegel** [1] and **Anton Wellstein** [1,*]

[1] Department of Oncology, Lombardi Comprehensive Cancer Center, Georgetown University, Washington, DC 20007, USA; anne.deslattesmays@jax.org (A.D.M.); mos6@georgetown.edu (M.S.); garrett.graham@georgetown.edu (G.G.); ms2465@georgetown.edu (M.S.); jmazarati@gmail.com (J.-B.M.); ariege01@georgetown.edu (A.R.)
[2] The Jackson Laboratory, Farmington, CT 06032, USA
[3] Pacific Biosciences, Menlo Park, CA 94025, USA; etseng@pacificbiosciences.com (E.T.); pbaybayan@pacificbiosciences.com (P.B.)
[4] Icahn School of Medicine at Mount Sinai, Institute for Genomics and Multi-scale Biology, New York, NY 10029, USA; robert.sebra@mssm.edu
[5] Biomedical Center, National Reference Laboratory, Kigali, Rwanda
* Correspondence: anton.wellstein@georgetown.edu; Tel.: +1-202-687-3672

Received: 19 February 2019; Accepted: 22 March 2019; Published: 27 March 2019

Abstract: Hematopoietic cells are continuously replenished from progenitor cells that reside in the bone marrow. To evaluate molecular changes during this process, we analyzed the transcriptomes of freshly harvested human bone marrow progenitor (lineage-negative) and differentiated (lineage-positive) cells by single-molecule real-time (SMRT) full-length RNA-sequencing. This analysis revealed a ~5-fold higher number of transcript isoforms than previously detected and showed a distinct composition of individual transcript isoforms characteristic for bone marrow subpopulations. A detailed analysis of messenger RNA (mRNA) isoforms transcribed from the *ANXA1* and *EEF1A1* loci confirmed their distinct composition. The expression of proteins predicted from the transcriptome analysis was evaluated by mass spectrometry and validated previously unknown protein isoforms predicted e.g., for *EEF1A1*. These protein isoforms distinguished the lineage negative cell population from the lineage positive cell population. Finally, transcript isoforms expressed from paralogous gene loci (e.g., *CFD*, *GATA2*, *HLA-A*, *B*, and *C*) also distinguished cell subpopulations but were only detectable by full-length RNA sequencing. Thus, qualitatively distinct transcript isoforms from individual genomic loci separate bone marrow cell subpopulations indicating complex transcriptional regulation and protein isoform generation during hematopoiesis.

Keywords: full length RNAseq; mRNA isoforms; protein isoforms; bone marrow cell subpopulations

1. Introduction

Alternative splicing of pre-messenger RNA (mRNA) generates multiple transcript isoforms from a single gene that can code for proteins with distinct functions. Examples include distinct ligand recognition of the b- versus c-isoforms of fibroblast growth factor (FGF) receptors [1], the pro- versus anti-apoptotic activity of BCL-X [2], and FAS isoforms [3], or distinct hormone sensitivity due to long and short isoforms of a transcriptional co-activator, AIB1 [4] (reviewed in References [5–7]). Indeed, for >90% of human genes, an average of five transcript isoforms are predicted, suggesting a challenging complexity of the protein-coding transcriptome [8]. Unbiased methods such as RNA sequencing are crucial for unraveling the transcriptome complexity in different tissues or cells under diverse physiologic or pathologic conditions [9–11].

Bone marrow continuously replenishes differentiated blood cells from a small pool of stem cells. Studies of this process have provided many of the concepts of tissue regeneration from resident stem cells [12–14]. Here, we compared the transcriptomes of progenitor and differentiated bone marrow cell populations using single-molecule real-time (SMRT) full-length RNA-seq of unfragmented complementary DNAs (cDNAs) abbreviated here as "full length RNA-seq" (Pacific Biosciences, Menlo Park, CA, USA; Iso-Seq). For this, we isolated progenitor (lineage-negative) and differentiated (lineage-positive) cell populations from intact, freshly harvested human bone marrow tissues. We identified a multitude of novel transcript isoforms whose composition distinguished progenitor and differentiated cell populations at most of the single genomic loci interrogated. These differences in transcript composition were not uncovered by a de novo analysis of conventional, short-read RNA-seq of fragmented cDNAs that was run in parallel. We confirmed translation of transcripts by mass spectrometry of proteins extracted from the bone marrow and identified novel protein isoforms predicted from the transcript analysis. To our knowledge, this is the first study that performs full-length RNA sequencing on segregated hematopoietic cell subpopulations followed by mass spectrometry validation. We conclude that bone marrow cell subpopulations are distinguishable at the single gene level by qualitative differences in transcript isoform composition suggesting more complex transcriptome regulation during hematopoiesis than previously described.

2. Methods

Healthy Bone Marrow Donors—The study was reviewed and considered as "exempt" by the Institutional Review Board of Georgetown University (IRB # 2002-022). All methods were carried out in accordance with relevant guidelines and regulations. Freshly harvested bone marrow tissues were collected from discarded healthy human bone marrow collection filters that had been de-identified. cDNA libraries for the short-read RNA-seq and the full-length RNA-seq were isolated from two separate healthy donors. From one donor, three cDNA library preparations were generated for total bone marrow, lineage-negative, and lineage-positive cells. Full-length RNA-seq of the samples was done at Pacific Biosciences. From an additional donor, a cDNA library was generated from lineage-negative cells. This was sequenced at the Mt. Sinai Sequencing Facility. Independent sequencing results from the same donor and from the additional donor were used for validation of the full-length RNA-seq results (see Figure 1b).

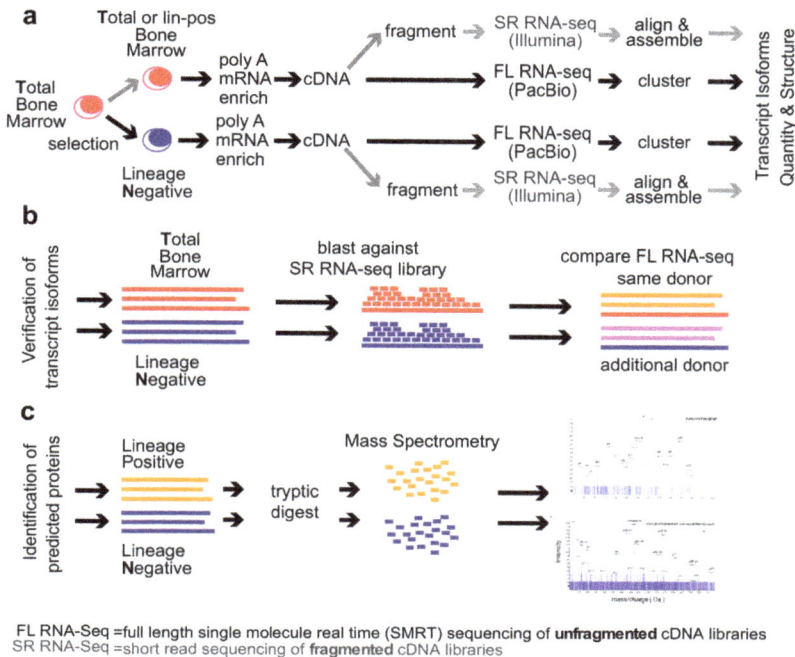

FL RNA-Seq =full length single molecule real time (SMRT) sequencing of **unfragmented** cDNA libraries
SR RNA-Seq =short read sequencing of **fragmented** cDNA libraries

Figure 1. Graphical abstract of the Transcriptome analysis of human bone marrow (BM) cell populations. (**a**) Flow of experiments and analyses. Poly(A)$^+$ RNA was isolated from Total (T, red) or lineage-negative BM cell populations (N, blue). Unfragmented, full-length complementary DNA (cDNA) libraries were subjected to single-molecule real-time (SMRT) RNA-seq (PacBio platform) or conventional short-read RNA-seq of fragmented cDNAs at 20 million (T) or 100 million (N) read depth (Illumina). Full-length RNA-seq data were processed using the ToFU platform. Illumina reads were first aligned and then assembled using the Tuxedo suite. The efficiency of the double selection for lineage-negative cells used here was confirmed by comparison of the abundance of standard markers of differentiated cells: CD14 = 6:1; CD16b = 25:1; CD24 = 109:1; CD45 = 11:1; CD66b = 16:1 expression ratio of lineage-positive to lineage-negative cells. (**b**) Validation of transcript isoforms identified by full-length RNA-seq. A BLAST-able library was generated from the short-read RNA-seq and used to align the isoforms identified by full-length RNA-seq. A separate full-length RNA-seq of an independent sample from the same donor, and a sample from a different donor were used for comparison. (**c**) Identification of proteins predicted by full-length RNA-seq using mass spectrometry of bone marrow cell extracts.

Healthy bone marrow cells—mononuclear cells were isolated by Ficoll gradient centrifugation. In order to select for lineage-negative cells, bone marrow mononuclear cells were incubated with an antibody cocktail containing antibodies against CD2, CD3, CD5, CD11b, CD11C, CD14, CD16, CD19, CD24, CD61, CD66b, and Glycophorin A (Stemcell Technologies, Vancouver, British Columbia, Canada). Lineage-positive cells bound to the antibodies were removed by magnetic beads and lineage-negative cells obtained from the flow-through. To increase purity, lineage-negative cells were enriched two times.

Short-read RNA-seq: Sequencing of fragmented cDNAs—Total RNA was submitted to Otogenetics Corporation (Norcross, GA, USA) for RNA-seq. Briefly, the integrity and purity of total RNA were assessed using Agilent Bioanalyzer by OD260/280 ratio. Altogether, 5 µg of total RNA was subjected to rRNA depletion using the RiboZero Human/Mouse/Rat kit (Epicentre Biotechnologies, Madison, WI, USA). cDNA was generated from the depleted RNA using random hexamers or custom primers and

Superscript III (Life Technologies, Carlsbad, CA, USA, catalog# 18080093). The resulting cDNA was purified and fragmented using a Covaris fragmentation kit (Covaris, Inc., Woburn, MA, USA), profiled using an Agilent Bioanalyzer, and Illumina libraries were prepared using NEBNext reagents (New England Biolabs, Ipswich, MA, USA). The quality, quantity, and the size distribution of the Illumina libraries were determined using an Agilent Bioanalyzer 2100. The libraries were then submitted for Illumina HiSeq2000 sequencing. Paired-end 90 or 100 nucleotide reads were generated and checked for data quality using FASTQC (Babraham Institute, Cambridge, UK), and DNAnexus (DNAnexus, Inc, Moutain View, CA, USA) was used on the platform provided by the Center for Biotechnology and Computational Biology (University of Maryland, Baltimore, MD, USA) [15]. A total of 159,043,023 non-strand-specific paired-end reads were collected from the total (T) and 35,126,712 strand-specific paired-end reads from the lineage-negative (N) bone marrow sample; 56.6% (T) and 51.3% (N) of the reads were mapped as a unique sequence using Tophat 2.

Full-length RNA-seq—total RNA was submitted to Pacific Biosciences (Menlo Park, CA, USA) or Icahn School of Medicine at Mount Sinai (New York, NY, USA). The integrity and purity of total RNA were assessed using Agilent Bioanalyzer and OD260/280 prior to submission. Full-length cDNA synthesis was done from polyA RNA using Clontech SMARTer PCR cDNA synthesis kit (Clontech Laboratories, Moutain View, CA, USA; [16]). Libraries were prepared after size selection of cDNAs into bins that contain 1–2 kb, 2–3 kb, and >3 kb cDNAs by the BluePippin size selection protocol (Sage Science, Beverly, MA, USA). These fractions were converted to single-molecule real-time (SMRT) libraries followed by SMRT sequencing. A total of 17 SMRT cells (7 cells 1–2 kb, 5 cells 2–3 kb, 5 cells 3–6 kb) were used to sequence the total bone marrow cell population, generating 234,078 single-molecule reads uniquely associated with non-redundant isoforms; 12 SMRT cells were used to sequence the lineage-negative population (5 cells 1–2 kb, 5 cells 2–3 kb, 2 cells 3–6 kb) generating 231,960 single-molecule reads uniquely associated with non-redundant isoforms Further, 8 SMRT Cells were used to sequence the lineage-negative population of another donor (4 cells <2 kb, 4 cells >2 kb) generating 195,614 single-molecule reads uniquely associated with non-redundant isoforms. Finally, 6 SMRT cells were used to sequence an alternative sample from the same donor, this was a lineage-positive population (2 cells 1–2 k, 2 cells 2–3 k, 2 cells 3–6 k) generating 74,434 single-molecule reads uniquely associated with non-redundant isoforms (see below, ToFU).

Mass Spectrometry Analysis of Proteins by Nano LC-MS/MS—proteins were extracted using 0.1% Rapigest (Waters Inc., Milford, MA, USA) in 25 mM ammonium bicarbonate Extracted proteins were reduced with 5 mM DTT for 60 min at 60 °C and alkylated with 15 mM iodoacetamide for 30 min in the dark. Trypsin (Promega, Madison, WI, USA) digestion (2.5 ng/μL) was carried out at 37 °C in Barocycler NEP2320 (Pressure BioSciences, Easton, MA, USA) for 1 h at 37 °C and then vacuum dried in Speed-vac (Labconco, Kansas City, MO, USA).

Tryptic peptides were analyzed on a NanoAcquity UPLC (Waters) by RP chromatography on a Symmetry C18 (3 μm, 180 μm, 20 mm) trap column and UPLC capillary column (BEH 300 Å, 1.7 μm, 150 mm × 0.75 μm) (Waters) interfaced with 5600 TripleTOF (AB Sciex, Framingham, MA, USA). Separation was achieved by a 250 min gradient elution with acetonitrile (ACN) containing 0.1% formic acid. The chromatographic method was composed of 5 min trapping step using 2% ACN at 15 μL/min and chromatographic separation at 0.4 μL/min as follows: Starting conditions 2% ACN; 1–180 min, 2%−60% ACN; 180−200 min, 60%−95% ACN; 200−220 min 95% ACN followed by equilibration 2% ACN for an additional 30 min. For all runs, 5 μL of sample were injected directly after enzymatic digestion. Analysis was conducted using an Information Dependent Acquisition (IDA) work flow with one full scan (400–1500 m/z) and 50 MS/MS fragmentations of major multiply charged precursor ions with rolling collision energy. Mass spectra were recorded in the MS range of 400–1500 m/z and MS/MS spectra in the range of 100−1800 m/z with resolution of 30,000 and mass accuracy up to 2 ppm using the following experimental parameters: Declustering potential, 80 V; curtain gas, 15; ion spray voltage, 2300 V; ion source gas 1, 20; interface heater, 180 °C; entrance potential, 10 V; collision exit potential, 11 V; exclusion time, 5 s; collision energy was set automatically according to m/z of the precursor (rolling collision energy). Data were processed using ProteinPilot 4.0 software (Sciex, Framingham, MA, USA) with a false discovery rate (FDR) of 1% [17,18]. Read-outs of the analysis are in Figure S4. For targeted measurements, an inclusion parent mass list was created according to in-silico tryptic digest of interesting sequences.

Transcriptome Alignment and Assembly from Illumina Data—Illumina reads were trimmed using Trimmomatic with the default parameters and the reads then aligned and assembled according to the Tuxedo suite protocol as described in Reference [15]. The genome of reference used was GRCh37 (hg19). The genes.gtf from this reference was used to guide the read alignment during the Tophat 2 step and Cufflinks2 [15]. Bowtie 2 indices were used for the genome reference. All computation was performed using Amazon Web Services and through the use of Starcluster software to manage the boxes. A Sun Grid Engine was employed to run the tasks. Reads were trimmed by Trimmomatic with the default parameters.

The specific qsub command for the total bone marrow (T) alignment are provided in the Supplemental Experimental Procedures.

*ToFU (=**T**ranscript is**o**forms: **F**ull-length and **U**nassembled; also named Isoseq3)*—Reads obtained from the Pacific Biosciences RS II platform were run through the ToFU pipeline to obtain high-quality non-chimeric full-length reads [19]. The original python code wrapped several separate processes, permitting the software to be run on a high-performance computing cluster. This bioinformatic process begins with circular consensus reads and classifies the reads into full-length (5′ primer seen, polyA tail seen and 3′ primer seen) and non-full-length reads. Primers and the poly A/T tail sequences are then removed and the transcript strandedness determined. Consensus is used to correct random errors once transcripts are assembled into full-length similar clusters. Further consensus error correction is performed with the non-full-length transcripts using a function named Quiver to generate the final full-length error-corrected transcripts. These were collapsed to the longest transcript and their abundance information obtained. A master ID was created to permit the comparison of transcript isoforms obtained from one sample population to another. The abundance information was then converted to Transcripts per Million according to the specifications of Li et al. [20]. Custom python and R Scripts were used to perform the analysis. These are available upon request.

To correct for random sequencing errors the ToFU generates high-quality fasta sequences with high accuracy due to the circular consensus reads, which are the input to the ToFU algorithm and do not suffer from the raw read error rate—raw read errors on the Pacific Biosciences platform are corrected through consensus due do the barbell primers used to construct the libraries resulting in a single molecule being read about 10 times by the polymerase in a typical library, thus correcting random errors. ToFU performs an inter-well consensus that involves a multi-step process of classifying reads as full length or not by determining the presence of both the 5′ and the 3′ primers, polyA tail and if it is chimeric. The non-full-length reads are used for further confirmation. Here we used only high-quality reads (>99%) to create our final consensus transcripts.

Conversion of FPKMs (=Fragments Per Kilobase of transcript per Million mapped reads) to transcripts per million *(TPMs)*—the conversion formula used is:

$$\text{TPM} = \text{FPKM}/(\text{sum of the FPKM over all genes/transcripts}) * 10^6 \qquad (1)$$

Conversion of ToFU abundance to TPMs—the conversion formula was generated according to Li et al. [20]. Abundance was obtained as output from ToFU and was translated to transcripts per million (TPM). Code is available upon request.

Transcript quality assessment by MatchAnnot—MatchAnnot is a python script that compares aligned full-length SMRT RNA-sequencing transcripts to existing annotation files. By starting with a transcript annotation file, MatchAnnot compares the full-length transcript and identifies a transcript within the genomic locus that provides the best match. Proceeding with this transcript, MatchAnnot then determines if there are skipped exons, alternative acceptor, and donor sites, etc. Comparisons are made using the transcript annotation file and the aligned input file (sorted SAM format). Start and end coordinate numbers are used to determine the details of the previously unknown transcript. Scores of 0 and 1 are considered poor matches. Scores of 2 or greater are considered as acceptable transcripts with viable alternative splicing annotation.

Confirmation of novel transcript isoforms with blast—to confirm novel transcripts isoforms, two BLAST-able databases were prepared separately for the total bone marrow (total.bm.non.ss) and lineage-negative short-read RNA sequences (lin.neg.ss) using the example script provided in Supplemental Experimental Procedures. For all reads where there was a gap, these were then compared with the full-length transcripts obtained from the alternative sample from the same donor (lineage-positive) and an alternative donor sample (lineage-negative) to arrive at the coverage numbers reported in Table S6.

Open Read Frame (ORF) Prediction—ORFs were generated using the ANGEL software publically available at GitHub (San Francisco, CA, USA), and through the use of SerialCloner 2.6.1. Franck Perez [SerialBasics]. ORFs accepted were the first ORFs and not necessarily the largest. In some cases, both the first and the largest ORF were included and designated with letters a, b, etc., appended to the end of the name assigned.

Multiple Sequence Alignment—multiple sequence alignment was done using Clustal Omega available at EBI (European Bioinformatics Institute, Hinxton, UK).

Sequence Alignment Editing—the sequence alignments were edited using BioEdit version 7.2.5. This Sequence Alignment Editor written for the windows environment was run on OSX Yosemite on a Mac Book Pro through the use of the wine version 1.6.2, a windows emulator available for download, and installation through Home Brew version 0.9.5.

Naming Convention—the names generated by the cDNA Primer software were used to create the nomenclature that relates to the deposited transcript structures as well as the uniprot deposited protein isoforms.

Quantitation of Abundance—all figures were generated through custom R scripts permitting alignment of abundance with isoforms. The full-length sequence reads were aligned to the hg19 reference genome using gsnap. The fragmented sequence reads were aligned to the hg19 reference using Tophat 2 and the aligned and paired reads were assembled using Cufflinks 2 and genes.gtf also from hg19 annotation as a guide [15]. Quantitation was as reported by Cufflinks in FPKMs and these were transformed to TPMs according the above formula. Data structures using Bioconductor packages GRanges were used to unify the results. All figures generated in R were edited within Adobe Illustrator.

Code availability—custom code is available upon request.

3. Results

We compared the total human bone marrow cell population (T) dominated by differentiated cells with the small (<1%) subpopulation of lineage-negative progenitor cells (N) using single-molecule real-time (SMRT) sequencing of unfragmented cDNAs (abbreviated here as "full-length RNA-seq"; Figure 1; [19]). Samples were also analyzed at 20 and 100 million read depths by conventional RNA-seq, a method that relies on the computational assembly of transcripts from short-read sequences of fragmented cDNAs [15,21]. We first focus on the analysis of two representative genes that are abundantly expressed in hematopoietic cells (*EEF1A1* and *ANXA1*). Utilizing complete transcriptomes generated either by full-length or short-read RNA-seq, we then describe the composite results and provide a comparison with protein fragments identified by mass spectrometry. The overall experimental flow is shown in Figure 1.

Distinct novel transcript isoforms of EEF1A1 detected in bone marrow cell subpopulations by full-length RNA-seq—eukaryotic translation elongation factor 1 α 1 (*EEF1A1*) is a highly abundant, conserved protein that delivers aminoacyl-tRNAs to the ribosome during protein synthesis but has also been found to contribute to additional cellular functions [22]. The *EEF1A1* gene spans 5.2 kb on chromosome 6q13 and results in a 3.5 kb transcript that contains six protein coding exons (RefSeq HG19; Figure 2a).

Short-read RNA-seq of total and lineage-negative bone marrow populations matched to the known reference transcripts translating for the canonical open reading frame (ST1-C, ST2-C*, in red for the Total; SN1-C, in blue for the lineage-negative; Figure 2b). One of the transcript isoforms, however, did not include the long 3′ UTR (ST2-C*) and one novel isoform in lineage-negative cells (SN2-3) skipped four coding exons due to an alternative splice acceptor and did not contain the long 3′ UTR in the RefSeq data base.

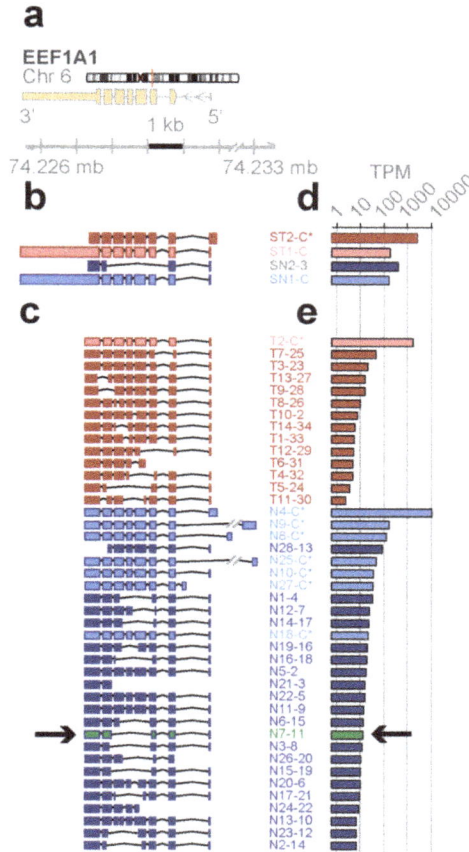

Figure 2. Messenger RNA (mRNA) isoforms of the *EEF1A1* gene. (**a**) Reference gene model from hg19. Arrows indicate direction of transcription. (**b–e**) Isoforms and abundances (TPM, transcripts per million reads) discovered in lineage-neg (blue) and total BM (red) cell populations by conventional short-read RNA-seq (**b,d**) or by full-length RNA-seq (**c,e**). Open reading frames (ORFs) and a novel protein isoform (arrows) that was confirmed by mass spectrometry for a unique peptide are shown in Figure 3. MatchAnnot confirmed the quality of the transcripts against gencode v19 (details in Methods). Abbreviations: S, short-read RNA-seq; C, canonical transcript and open reading frame (ORF); C*, non-canonical transcript isoforms with canonical ORF. ID#s of the isoforms are from the identifiers generated by the sequencing method. Ensembl IDs are included in Table S8, NCBI Accession numbers in Table S12.

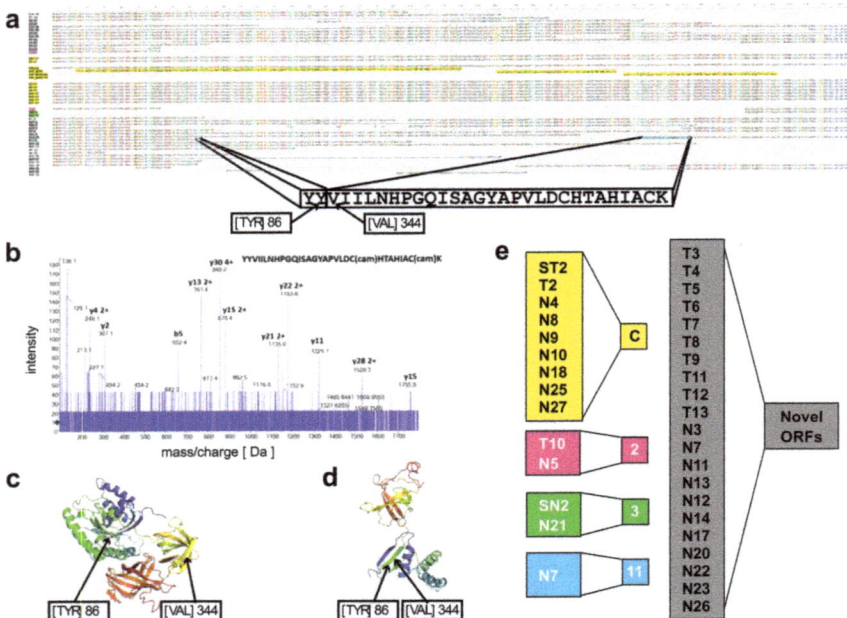

Figure 3. Proteins predicted from the transcript isoforms identified for *EEF1A1*. (**a**) Amino acid sequence alignments including the transcript isoform identifiers (see Figure 2b,c). The protein predicted from the N7-11 transcript isoform (highlighted in blue) joins Y86 and V344 in the canonical protein. The amino acid sequence of a predicted signature tryptic fragment peptide is shown. (**b**) Manually assigned fragmentation spectra from targeted analysis by Nano LC-MS/MS of proteins extracted from lineage-negative bone marrow cells with the sequence read of the N7-24 signature peptide; cam, carbamidomethylation. (**c,d**) Protein structure models were generated in Phyre2 [23]. The predicted structure of the canonical *EEF1A1* protein P68104 (**c**) and of the novel N7 protein (**d**) are shown. The c1g7ca template of P68104 used for the model covered 90% of the amino acids in the N7 isoform. (**e**) Higher magnification of transcript isoform identifiers that code for the canonical protein (yellow background). C, canonical RefSeq derived transcripts. C*, previously unknown transcript isoforms that code for the canonical protein. Transcript isoforms SN2-3 and N21-3 predict the same protein (green background).

In contrast, the *EEF1A1* transcript isoforms obtained from full-length RNA-seq (Figure 2c) showed eight transcript isoforms each translating for the same canonical protein isoform (P68104; see Figure 3a). In total, seven of these isoforms were found in the lineage-negative (N4-C*, N9-C*, etc.,) and one was found in the total bone marrow cell population (T2-C*). An additional 34 previously unknown transcript isoforms were identified, 14 of which were found in the total and 28 in the lineage-negative cell populations. Exon-skipping was seen in 22 (e.g., T9-28), exon-splitting in 2 (T8-26, N13-10), and alternative donor/acceptor sites in 30 transcript isoforms (e.g., T7-25). No transcripts were found that contained the long 3′ UTR annotated in RefSeq (Figure 2c). The estimated abundance of transcript isoforms is shown in Figure 2d,e.

A novel protein of EEF1A1 predicted from full length RNA-seq confirmed by mass spectrometry—open reading frames predicted from the transcript isoforms of *EEF1A1* were aligned to the canonical protein and overlaid with its 3D structure (Figure 3a,c). Of the 21 novel protein isoforms predicted (Figure 3a,e), the N7 protein contained a unique tryptic peptide fragment that was distinct from the canonical EEF1A1 protein and thus potentially detectable by mass spectrometry. The N7 transcript is found only in the lineage-negative cell population at low abundance (21 TPM; see Figure 2e) and is predicted to code for a 205 aa protein that lacks the central 258 amino acids of the EEF1A1 protein joining Y86 with V344 (Figure 3a). A unique tryptic peptide fragment spanning Y86 to V344 was detected by mass spectrometry analysis of proteins extracted from lineage-negative cells (Figure 3b) and thus confirms that the N7-11 transcript is translated into a protein expressed at sufficient levels detectable by mass spectrometry. It is noteworthy that the protein segment containing the Y86-V344 junction also provides a distinct epitope that is not present in the canonical protein.

Novel ANXA1 transcript isoform composition uncovered by full length RNA-seq distinguish bone marrow cell subpopulations—annexins are known as organizers of membrane dynamics and include binding proteins for endocytosis, exocytosis, and other localization functions [24]. The *ANXA1* gene spans 18 kb on chromosome 9q21 with 12 coding exons and results in an approximately 1.5 kb transcript (Figure 4a). With the conventional short-read RNA-seq and computational transcript reconstruction, only the canonical *ANXA1* transcript was found in both total bone marrow and lineage-negative cell populations (Figure 4b). That is, there were no distinguishing transcript isoforms separating the total bone marrow from the lineage-negative cell population. In contrast, with full-length RNA-seq (Figure 4c), 38 transcript isoforms were identified; amongst these were transcript isoforms specific to the total bone marrow cell population (T7, T14) and transcript isoforms specific to the lineage-negative cell population (N2, N3, N4, N8). While there were shared transcript isoforms between the two cell populations, there were transcript isoforms found only in one population and not in the other—21 were found in the total and 17 were found in the lineage-negative cell population. Two of these novel isoforms predict the canonical protein, P04083 (T10-C*, N12-C*), while others contain distinct open reading frames generated by exon-skipping (25 isoforms; e.g., T16-12, N7-12), alternative donor and acceptor sites (11 isoforms; e.g., T19-6, N10-2), and intron retentions (9 isoforms; e.g., T6-14, N11-13).

Proteins predicted from the ANXA1 transcript isoforms—the canonical *ANXA1* protein contains four repeat regions (r1 to r4) of approximately 70 amino acids, each with a motif for calcium binding. These sequences are highlighted in yellow and the alignment of the predicted ORFs from transcript isoforms in total and lineage-negative bone marrow cells shows the overlaps with the canonical protein and with each other (Figure S1a,b). Mass spec analysis of the proteins extracted from bone marrow confirmed the presence of matching peptides in both cell populations (red highlights, total; blue highlights, lineage-negative cells; Figure S1a).

Four different transcript isoforms (T1-C, T10-C*, N1-C, and N12-C*) code for the canonical *ANXA1* protein. The T10-C* and N12-C* transcripts are structurally different from T1-C and N1-C, each containing a novel exon in the 5' UTR (see Figure 4c). ORF 2 matches with the r4 repeat and is contained in seven different transcript isoforms whilst ORF 12, 13, and 14 match with the r1 to r3 repeats. These transcripts were detected in both cell populations. Additionally, ORF 3 (T14, T7), ORF 23 (N4, N8), and ORF 26 (N2, N3) are derived from two different transcript isoforms but found in only one of the cell populations (Figure S1a,c). Finally, five novel protein isoforms due to exon-skipping are predicted from the transcript isoforms in total and lineage-negative cells (T13, T15, T19, N13, and N9). The identification of groups of isoforms seen only in one cell population but not the other shows that isoform composition can be sufficient to distinguish bone marrow cell subpopulations.

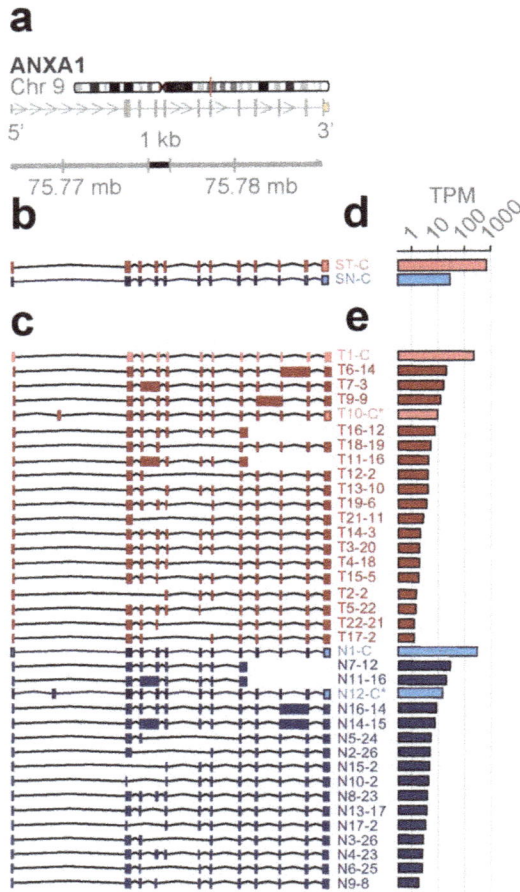

Figure 4. mRNA isoforms of the *ANXA1* gene. (**a**) Reference gene model from hg19. Arrows indicate direction of transcription. (**b–e**) Results for lineage-neg (blue) and total BM (red) from short-read RNA-seq (**b,d**) or full-length RNA-seq (**c,e**). Isoforms and abundances (TPM, transcripts per million reads) discovered in lineage-neg (blue) and total BM (red) population by short-read (**b,d**) or full-length RNA-seq (**c,e**). S, short-read RNA-seq; C, canonical transcript and open reading frame (ORF); C*, non-canonical transcript isoform with canonical ORF; ORFs are shown in Figure S1. ID#s of the isoforms are from the identifiers generated by the sequencing methods. Ensembl IDs are included in Table S9, NCBI Accession numbers in Table S12.

High complexity of the transcriptome revealed by full-length RNA-seq—as described above for *ANXA1* and *EEF1A1*, full-length RNA-seq of bone marrow cell populations revealed an up to 10-fold higher number of transcript isoforms than found by short-read RNA-seq. To investigate the universality of isoform complexity, we evaluated genes having <6 coding exons (*UBC, KLF6, LYZ, SAT1*) and found that RNA-seq detected 1 to 4 isoforms while full-length RNA-seq identified 12 to 36 isoforms. In a repeat sample, we identified as many as 15 isoforms with full-length RNA-seq, and in a sample from an additional donor we identified a maximum number of 48 isoforms for these genomic loci. We then considered loci with >6 coding exons. Short-read RNA-seq yielded a similar number of isoforms as it had with <6 exons, i.e., 1 to 3 isoforms compared with 31 to 43 transcript isoforms identified by full-length RNA-seq (Figure 5a). Similarly, in a repeat sample we identified 4 to 20 isoforms and from an additional donor 3 to 43 transcript isoforms at the genomic loci with >6 coding exons (EEF1A1, GLUL, HLA-E, CD74, PKM, and ANXA1).

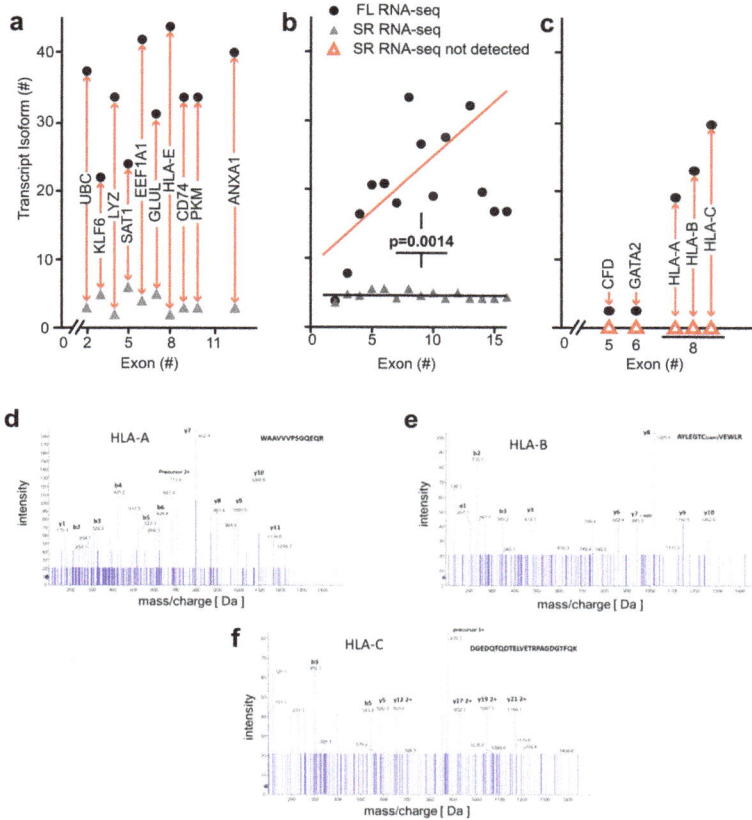

Figure 5. Transcript isoform detection for genomic loci of different complexity and redundancy. (**a–c**) The number of transcript isoforms detected (y-axes) by full-length (FL; filled circles) and short-read (SR; triangles) RNA-seq is shown relative to the number of canonical exons (x-axes; hg19 gene annotation). (**a**) Representative genes with 3 to 13 canonical exons. Full gene names and isoform numbers are provided in Table S3. (**b**) Comparison of the mean of the transcript isoforms for the five most abundant genes with 2 to 16 exons from the analysis of all bone marrow cell populations ($p = 0.0014$; Chi-sq. for trend short-read (SR) RNA-seq versus full-length (FL) RNA-seq). The numbers of transcript isoforms for each of the subgroups are in Table S5a. (**c**) Transcripts and isoforms identified only by full-length RNA-seq. Gene names and numerical values are in Table S4. (**d–f**) Manually assigned fragmentation spectra for peptides matching with HLA-A, -B, and -C transcripts. Peptides were detected using non-targeted shotgun proteomics analysis of proteins extracted from lineage-negative bone marrow cells. Detection of the peptides confirms expression of HLA-A, -B, and -C.

We extended the analysis by arranging genes by mean exon number and identified the loci with the top five transcript isoform counts in each bin (Tables S1 and S2). Much to our surprise, the number of transcript isoforms identified by short-read RNA-seq plateaued at four isoforms irrespective of the number of exons in a given gene (Figure 5b). In contrast, full-length RNA-seq showed an increase in transcript isoform number with increasing complexity of genomic loci as indicated by the number of canonical exons ($p = 0.0014$). It is worth noting that an increase in sequencing depth for short-read RNA-seq from 20 to 100 million reads did not impact this maximum significantly ($p > 0.05$; Table S1), an observation that matches with earlier predictions [25]. Thus, our analysis supports the notion that the complexity of the transcriptome will be underestimated by short-read RNA-seq regardless of the complexity of genomic loci evaluated.

Matching of newly discovered and known transcript isoforms—we compared the transcript isoforms found in the original runs as well as replicate samples from an additional donor with known transcript isoforms using matchAnnot (see Experimental Procedures); 26% to 58% of aligned reads showed the top matchAnnot score of >5 relative to known transcript isoforms deposited in the data base (Table S7). Scores of ≥2, i.e., an acceptable match with some mismatched splicing, was seen for 88% to 98% of the transcripts detected. The qualitative differences in the transcript composition uncovered by full-length RNA-seq became apparent in this overall comparison where all transcript can be assigned to a genomic locus but only 39% of currently known transcript isoform compositions match with the isoforms described here (Table S7). As shown above for a set of genes, the transcript isoform compositions enable one to distinguish lineage-positive from lineage-negative cell populations.

Detection of novel transcripts of paralogous genes—paralogous genes with conserved sequences pose a particular challenge to detection by short-read RNA-seq and we evaluated the potential of full-length RNA-seq to gain insights into genomic loci containing gene paralogs. Full-length RNA-seq identified transcripts for *CFD*, *GATA2*, *HLA-A*, *-B*, and *-C* that were missed by short-read RNA-seq, as shown in Figure 5c. The inability to detect or assign transcripts for these loci with short-read RNA-seq may be explained by the paralogous nature of the genes involved: *CFD* is located on chr 19 with *AZU1*, *PRTN3*, and *ELANE*. These four genes rank second in the list of regions of homozygosity cold spots on human autosomes and this genomic region underwent rapid Alu-mediated expansion during primate evolution creating the largest known microRNA gene cluster of the human genome [26]. Perhaps not surprisingly, transcript from a repetitive region of this kind was particularly problematic to resolve with short-read RNA-seq. Additionally, the ELANE and CFD proteins are 78% homologous (Figure S2; canonical proteins). Short-read RNA-seq uncovered transcripts from three of these four genes, missing *CFD*. That is not surprising, because blastn analysis matched *CFD* fragment sequences to *ELANE*. Using full-length RNA-seq, we were able to identify the transcripts as well as unique isoforms for each of the genes in these regions of homozygosity. Similarly, *HLA-A*, *HLA-B*, and *HLA-C* are paralogs with >80% identity that cannot be mapped appropriately and detected by short-read RNA-seq (Figure 5c and Supplemental Figure S3a). These data suggest that the presence of transcripts from paralogs adds to the complexity of alignments and obfuscate transcript reconstruction from short reads as well as the estimate of transcript abundance.

Mass spectrometry identification of peptides predicted from the transcriptome analysis—to assess the biological relevance of the transcript isoforms found and their predicted open reading frames, we identified detectable proteins in the bone marrow cell populations by mass spectrometry. We assessed the detection of matching peptides for a range of genomic loci that contain between 2 and 16 exons and were also associated with the highest number of transcript isoforms. Peptides were confirmed for 52 of the 150 transcripts depicted in Figure 5b (details in Table S2).

As mentioned above, *HLA-A*, *-B*, and *-C* transcript isoforms were only detected by full-length RNA-seq (Figure 5c). It is noteworthy that mass spectrometry identified distinct peptides that were predicted for the *HLA-A*, *B*, and *C* transcript isoforms. The mass spec readings for three of the peptides are shown in Figure 5d–f and the alignment of these sequences in Figure S3b. This supports the significance of the detection of these *HLA* transcripts and of the ORFs derived from the full-length RNA-seq.

Information gained on transcript isoforms and their abundance—for full-length RNA-seq, 56% of the transcripts mapped to loci with four or more exons and 31% mapped to loci with eight or more exons. In contrast, short-read RNA-seq mapped only 13% of transcripts to loci with >4 and 5% with >8 exons (Table S5c). Thus, full-length RNA-seq provides a significant ($p < 0.0001$) 4- to 6-fold gain in information. Given that over half of the detected transcripts are from multi-exon genes, the ability to span two exons with short reads may be inadequate to resolve a full-length transcript successfully without the addition of longer reads [21]. Also, ambiguous mapping of the short reads explains the high number of transcripts being mapped to genes with one or two exons (Table S5c). Our data also shows that short-read RNA-seq reaches a maximum of approximately four transcript isoform even for complex loci [8] whilst full-length RNA-seq shows a significant increase in transcript isoforms with increasing complexity of genomic loci (Figure 5b). As described in the more detailed analyses of *ANXA1* and *EEF1A1* above and an overview of genomic loci interrogated, full-length RNA-seq reveals complex posttranscriptional processing (Figure 6; Table S5).

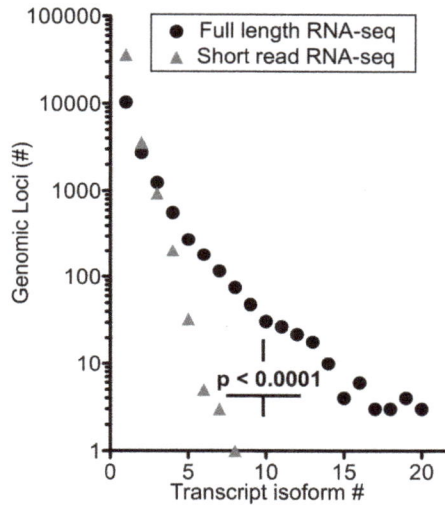

Figure 6. Number of transcript isoforms mapped to distinct genomic loci. Comparison of short-read (SR) versus full-length (FL) RNA-seq, $p < 0.0001$; Chi-square for trend. The distribution of exon counts per transcript isoform and transcript isoform counts per genomic locus are provided in Table S5c,d.

Confirmation of novel isoforms—we further confirmed the novel isoforms obtained in full-length RNA-seq in several different ways. First, we created a BLAST-able database using trimmed raw unaligned short-read RNA-seq data from each of the lineage-negative and total bone marrow populations. Limiting the results to only those reads with a 100 base pair coverage, novel transcript isoforms were confirmed by ensuring gapless coverage. Second, novel transcript isoforms were confirmed against full-length RNA sequencing libraries prepared from two additional samples (biological replicate samples). From a combination of these two approaches we confirmed an average of 83% of the transcript isoforms (range 74% to 100%; Table S6). Third, we ran computational experiments with public data (Tables S10 and S11). From these analyses we found that the full-length RNA-seq mostly showed novel combinations of known junctions. 87% of all junctions for EEF1A1 have coverage from short-read data with 100% of the new transcript isoforms sharing an open reading frame with the known EEF1A1 protein. In total, 94% of all the junctions for ANXA1 have coverage from short-read data and 100% of the novel transcript isoforms share an open reading frame with known ANXA1 protein. Fourth, we assessed the transcription start site (TSS) matches with the 5′-end of the full-length RNA-seq. For EEF1A1, ~90% and for ANXA1, ~94% were confirmed, respectively (data and details in Tables S8 and S9). Fifth, we aligned the EEF1A1 and ANXA1 transcript isoforms described here with the EST database. We found that most of the isoforms showed 97% to 100% alignment identity with EST sequences with only a very few as low as 90% identity (Figures S5 and S6).

4. Discussion

Our transcriptome analysis of the human bone marrow progenitor and differentiated cell populations revealed a previously unknown complexity of the transcripts when analyzed by single-molecule real-time (SMRT) sequencing of unfragmented cDNAs, i.e., full-length RNA-seq. With a direct reading of full-length RNA sequences—rather than computational assembly from read fragments—open reading frames are predicted more reliably and a proteomics analysis for the respective predicted proteins showed that one third of them were present at high enough abundance to be detected by mass spectrometry. The full-length RNA-seq analysis described here is the most recent approach that relies on the long reads possible with the PacBio technology [27] and the use of full-length, non-fragmented cDNAs as templates. The template cDNAs are generated by the SMART (switching mechanism at 5′ end of RNA template) approach established earlier [16] and favors generation of full-length cDNAs that include the 5′-end of intact mRNA. This provides reliable assessments of open reading frames that start at the most 5′ translation initiation codon.

Numerous computational experiments were conducted exploring alternative methods, including using Trinity and Oasis to perform de novo assembly of the short reads alone, then combined with the full-length reads. These were experiments with varied success. Parts of transcripts would often be confirmed, but the process was both time consuming and computationally costly with unsatisfactory outcome. It is remarkable that the complexities of the transcriptome and hence distinction between subpopulations of bone marrow cells described above were confirmed but not identifiable de novo by the conventional, short-read RNA-seq of fragmented cDNAs. Indeed, short-read RNA-seq uses computational methods to reconstruct transcripts from the sequence fragments and can uncover novel exon-exon junctions. However, for acceptable performance, reconstruction requires a transcript isoform database to resolve the origin of isoforms. This will bias assignments of sequence fragments and limits discovery of isoforms of higher complexity [28]. Thus, with the sequence fragments provided by short-read RNA-seq it is difficult to uncover the composition of a complex transcriptome as is evident from the present study and was described before [29,30]. Greater sequencing depth is one potential remedy to consider and appears to improve the detection of complex transcripts only in *D. melanogaster* and *C. elegans* but not in *H. sapiens*. This is likely due to the lower number of transcript isoforms per genomic locus found in the invertebrates—fewer than 25% of genes in *C. elegans* and *D. melanogaster* give rise to more than two transcript isoforms, whereas human genes are currently annotated with an average of five transcript isoforms. Beyond this, short-read RNA-seq may identify too few isoforms due to ambiguous assignment of a given read to multiple loci. Altogether, this will impede the ability to detect novel isoforms [9,31,32].

The most striking examples that reveal the limitations of the short-read RNA-seq are paralogous gene loci where only full-length RNA-seq detected gene expression (Figure 5c). Mass spectrometry identified peptides in the bone marrow cell populations that match with the ORFs predicted from the full-length RNA-seq of the paralogous loci (Figure 5d–f and Figure S3b). Interestingly, when the full-length transcript sequences were used as templates to search for matches in the raw data from short-read RNA-seq, 11% and 57% were confirmed for *HLA-A* and *-C*, respectively (Table S6). These findings highlight the power as well as the limitations of the different approaches to RNA-seq.

An analysis of splicing complexity by gene size showed that the combination potential for exon splicing can increase exponentially [33]. The authors speculated that functional and evolutionary constraints are the reason why the number of transcripts thus far seen is less than the theoretical maximum. Our data indicate, however, that the number of transcripts reported to date may be limited by the short-read reconstruction approach. Full-length RNA-seq is able to reveal unique transcript isoform structures beyond those already discovered and allows for the unambiguous attribution of reads to defined genomic loci [11,30,33,34].

We conclude that single-molecule real-time full-length RNA-seq provides a map of the complexity of transcript isoform structures that eludes the transcriptome analysis by conventional short-read RNA-seq. In addition, open reading frames of proteins derived from these full-length transcript isoforms enables the comparison of the presence and absence of specific binding sites, domains, and regulatory regions in proteins that can be related to the cell population and physiologic or pathologic status [35]. In particular, the analysis of paralogous genes revealed the power of full-length RNA-seq to identify transcripts even for complex loci where short-read fragment sequencing failed. Transcriptome analysis by full-length RNA-seq provided a map of discrete transcript structures that can serve as qualitative markers for cell subpopulations in the bone marrow. This provides a qualitative expansion beyond the comparison of patterns based on the abundance of gene expression. Also, this indicates a more complex regulation of genes during hematopoiesis than previously appreciated.

Finally, as shown above, protein isoforms contain isoform-specific amino acid sequence junctions that represent epitopes specific for cell subpopulations in complex tissues. It is conceivable that pathologic regulation of splicing in diseases such as neurodegenerative disorders, metabolic syndrome, or cancer can generate neo-epitopes and elicit an immune response irrespective of genomic mutagenesis. Unraveling the complexity of the transcriptome complemented by analysis of resulting proteins thus may provide new diagnostic options and unique therapeutic targets.

Supplementary Materials: The following are available online at http://www.mdpi.com/2073-4425/10/4/253/s1, File S1: Supplemental Experimental Procedures; Table S1: Top five transcript isoform counts for genes with 2 to 16 exons, Table S2: Transcript ID, gene names, detection of peptides by mass spectrometry, Table S3: Number of exons, transcript isoform numbers (Trans #) for genes in Figure 5a, Table S4: Number of exons, transcript isoform numbers (Trans #) for genes in Figure 5c, Table S5: Distribution of exon counts and transcript isoform frequency, Table S6: Validation of the transcript isoforms, Table S7: MatchAnnot analysis of isoform alignments, Table S8: Supporting evidence for novel transcript isoforms of EEF1A1 using SQANTI2 and CAGE, Table S9: Supporting evidence for novel transcript isoforms of ANXA1 using SQANTI2 and CAGE, Table S10: Supporting evidence for exon-exon junctions in EEF1A1 with public RNAseq data, Table S11: Supporting evidence for exon-exon junctions in ANXA1 with public RNAseq data, Table S12: NCBI accession numbers.

Author Contributions: A.D.M. developed the analysis method, performed experiments, analyzed data, and wrote the paper. M.S., E.T., P.B., M.S., and J.-B.M. performed experiments, analyzed the associated data, and prepared figures and text for the paper. G.G assisted with method development, data analysis, and editing. R.S. assisted with method development and data analysis. A.W. and A.R. oversaw the research, analyzed data, and wrote the paper. All authors read and approved the manuscript for submission.

Funding: This work was supported in part by the National Institutes of Health Grants CA71508, CA51008 and CA177466 to A.W.; CA113477 to A.T.R. as well as internal funds from the Lombardi Cancer Center.

Acknowledgments: We are grateful to the leadership and staff of the bone marrow donor program at Georgetown University for their continued support. They saved discarded bone marrow collection filters for our retrieval of bone marrow debris of tissue fragments retained in the filters that are otherwise discarded as biohazardous waste. We also wanted to thank the anonymous donors of the bone marrow donor program. We acknowledge support by the Proteomics Shared Resources of the Lombardi Cancer Center.

Conflicts of Interest: Georgetown University filed a provisional patent application that is related to some aspects of this manuscript.

References

1. Goetz, R.; Mohammadi, M. Exploring mechanisms of FGF signalling through the lens of structural biology. *Nat. Rev. Mol. Cell Biol.* **2013**, *14*, 166–180. [PubMed]

2. Boise, L.H.; González-García, M.; Postema, C.E.; Ding, L.; Lindsten, T.; Turka, L.A.; Mao, X.; Nuñez, G.; Thompson, C.B. bcl-x. a bcl-2-related gene that functions as a dominant regulator of apoptotic cell death. *Cell* **1993**, *74*, 597–608. [PubMed]

3. Cheng, J.; Zhou, T.; Liu, C.; Shapiro, J.P.; Brauer, M.J.; Kiefer, M.C.; Barr, P.J.; Mountz, J.D. Protection from Fas-mediated apoptosis by a soluble form of the Fas molecule. *Science* **1994**, *263*, 1759–1762.

4. Reiter, R.; Wellstein, A.; Riegel, A.T. An isoform of the coactivator AIB1 that increases hormone and growth factor sensitivity is overexpressed in breast cancer. *J. Biol. Chem.* **2001**, *276*, 39736–39741.

5. Chen, J.; Weiss, W.A. Alternative splicing in cancer: Implications for biology and therapy. *Oncogene* **2015**, *34*, 1–14.

6. Bonnal, S.; Vigevani, L.; Valcárcel, J. The spliceosome as a target of novel antitumour drugs. *Nat. Rev. Drug Discov.* **2012**, *11*, 847–859.

7. Chen, M.; Manley, J.L. Mechanisms of alternative splicing regulation: Insights from molecular and genomics approaches. *Nat. Rev. Mol. Cell Biol.* **2009**, *10*, 741–754. [PubMed]

8. Pan, Q.; Shai, O.; Lee, L.J.; Frey, B.J.; Blencowe, B.J. Deep surveying of alternative splicing complexity in the human transcriptome by high-throughput sequencing. *Nat. Genet.* **2008**, *40*, 1413–1415. [PubMed]

9. Au, K.F.; Sebastiano, V.; Afshar, P.T.; Durruthy, J.D.; Lee, L.; Williams, B.A.; van Bakel, H.; Schadt, E.E.; Reijo-Pera, R.A.; Underwood, J.G.; et al. Characterization of the human ESC transcriptome by hybrid sequencing. *Proc. Natl. Acad. Sci. USA* **2013**, *110*, E4821–E4830. [PubMed]

10. Van Keuren-Jensen, K.; Keats, J.J.; Craig, D.W. Bringing RNA-seq closer to the clinic. *Nat. Biotechnol.* **2014**, *32*, 884–885. [PubMed]

11. Selvanathan, S.P.; Graham, G.T.; Erkizan, H.V.; Dirksen, U.; Natarajan, T.G.; Dakic, A.; Yu, S.; Liu, X.; Paulsen, M.T.; Ljungman, M.E.; et al. Oncogenic fusion protein EWS-FLI1 is a network hub that regulates alternative splicing. *Proc. Natl. Acad. Sci. USA* **2015**, *112*, E1307–E1316.

12. Pappenheim, A. Ueber entwickelung und ausbildung der erythroblasten. *Arch. Pathol. Anat. Physiol.* **1896**, *XXII*, 39.

13. Goodell, M.A.; Nguyen, H.; Shroyer, N. Somatic stem cell heterogeneity: Diversity in the blood, skin and intestinal stem cell compartments. *Nat. Rev. Mol. Cell Biol.* **2015**, *16*, 299–309. [PubMed]

14. Morrison, S.J.; Scadden, D.T. The bone marrow niche for haematopoietic stem cells. *Nature* **2014**, *505*, 327–334. [PubMed]

15. Trapnell, C.; Roberts, A.; Goff, L.; Pertea, G.; Kim, D.; Kelley, D.R.; Pimentel, H.; Salzberg, S.L.; Rinn, J.L.; Pachter, L. Differential gene and transcript expression analysis of RNA-seq experiments with TopHat and Cufflinks. *Nat. Protoc.* **2012**, *7*, 562–578.

16. Chenchik, A.; Zhu, Y.Y.; Diatchenko, L.; Li, R.; Hill, J.; Siebert, P.D. Generation and use of high-quality cDNA from small amounts of total RNA by SMART PCR. In *Gene Cloning and Analysis by RT-PCR*; Siebert, P.D., Larrick, J., Eds.; BioTechniques Books: Natick, MA, USA, 1998; pp. 305–319.

17. Shilov, I.V.; Seymour, S.L.; Patel, A.A.; Loboda, A.; Tang, W.H.; Keating, S.P.; Hunter, C.L.; Nuwaysir, L.M.; Schaeffer, D.A. The paragon algorithm, a next generation search engine that uses sequence temperature values and feature probabilities to identify peptides from tandem mass spectra. *Mol. Cell. Proteom.* **2007**, *6*, 1638–1655.

18. Tang, W.H.; Shilov, I.V.; Seymour, S.L. Nonlinear fitting method for determining local false discovery rates from decoy database searches. *J. Proteome Res.* **2008**, *7*, 3661–3667. [PubMed]

19. Gordon, S.P.; Tseng, E.; Salamov, A.; Zhang, J.; Meng, X.; Zhao, Z.; Kang, D.; Underwood, J.; Grigoriev, I.V.; Figueroa, M.; et al. Widespread polycistronic transcripts in fungi revealed by single-molecule mRNA sequencing. *PLoS ONE* **2015**, *10*, e0132628.

20. Li, B.; Ruotti, V.; Stewart, R.M.; Thomson, J.A.; Dewey, C.N. RNA-Seq gene expression estimation with read mapping uncertainty. *Bioinformatics* **2010**, *26*, 493–500.

21. Kim, D.; Pertea, G.; Trapnell, C.; Pimentel, H.; Kelley, R.; Salzberg, S.L. TopHat2: Accurate alignment of transcriptomes in the presence of insertions, deletions and gene fusions. *Genome Biol.* **2013**, *14*, R36.

22. Mateyak, M.K.; Kinzy, T.G. eEF1A: Thinking outside the ribosome. *J. Biol. Chem.* **2010**, *285*, 21209–21213.

23. Kelley, L.A.; Mezulis, S.; Yates, C.M.; Wass, M.N.; Sternberg, M.J.E. The Phyre2 web portal for protein modeling, prediction and analysis. *Nat. Protoc.* **2015**, *10*, 845–858.

24. Gerke, V.; Creutz, C.E.; Moss, S.E. Annexins: Linking Ca2+ signalling to membrane dynamics. *Nat. Rev. Mol. Cell Biol.* **2005**, *6*, 449–461. [PubMed]

25. Tarazona, S.; García-Alcalde, F.; Dopazo, J.; Ferrer, A.; Conesa, A. Differential expression in RNA-seq: A matter of depth. *Genome Res.* **2011**, *21*, 2213–2223.

26. Pemberton, T.J.; Absher, D.; Feldman, M.W.; Myers, R.M.; Rosenberg, N.A.; Li, J.Z. Genomic patterns of homozygosity in worldwide human populations. *Am. J. Hum. Genet.* **2012**, *91*, 275–292. [PubMed]

27. Rhoads, A.; Au, K.F. PacBio sequencing and its applications. *Genom. Proteom. Bioinform.* **2015**, *13*, 278–289.

28. Barrett, C.L.; DeBoever, C.; Jepsen, K.; Saenz, C.C.; Carson, D.A.; Frazer, K.A. Systematic transcriptome analysis reveals tumor-specific isoforms for ovarian cancer diagnosis and therapy. *Proc. Natl. Acad. Sci. USA* **2015**, *112*, E3050–E3057.

29. Su, Z.; Mason, C.E.; SEQC MAQC-III Consortium. A comprehensive assessment of RNA-seq accuracy, reproducibility and information content by the sequencing quality control consortium. *Nat. Biotechnol.* **2014**, *32*, 903–914.

30. Li, S.; Tighe, S.W.; Nicolet, C.M.; Grove, D.; Levy, S.; Farmerie, W.; Viale, A.; Wright, C.; Schweitzer, P.A.; Gao, Y.; et al. Multi-platform assessment of transcriptome profiling using RNA-seq in the ABRF next-generation sequencing study. *Nat. Biotechnol.* **2014**, *32*, 915–925.

31. Steijger, T.; Abril, J.F.; Engström, P.G.; Kokocinski, F.; Hubbard, T.J.; Guigo, R.; Harrow, J.; Bertone, P. RGASP consortium assessment of transcript reconstruction methods for RNA-seq. *Nat. Methods* **2013**, *10*, 1177–1184.

32. Engström, P.G.; Steijger, T.; Sipos, B.; Grant, G.R.; Kahles, A.; Rätsch, G.; Goldman, N.; Hubbard, T.J.; Harrow, J.; Guigo, R.; et al. RGASP Consortium Systematic evaluation of spliced alignment programs for RNA-seq data. *Nat. Methods* **2013**, *10*, 1185–1191. [PubMed]

33. Li, S.; Mason, C.E. The pivotal regulatory landscape of RNA modifications. *Annu. Rev. Genom. Hum. Genet.* **2014**, *15*, 127–150.

34. Kratz, A.; Carninci, P. The devil in the details of RNA-seq. *Nat. Biotechnol.* **2014**, *32*, 882–884. [PubMed]

35. Keren, H.; Lev-Maor, G.; Ast, G. Alternative splicing and evolution: Diversification, exon definition and function. *Nat. Rev. Genet.* **2010**, *11*, 345–355. [PubMed]

MDPI
St. Alban-Anlage 66
4052 Basel
Switzerland
Tel. +41 61 683 77 34
Fax +41 61 302 89 18
www.mdpi.com

Genes Editorial Office
E-mail: genes@mdpi.com
www.mdpi.com/journal/genes

www.ingramcontent.com/pod-product-compliance
Lightning Source LLC
Chambersburg PA
CBHW051911210326
41597CB00033B/6114